PESTS

PESTS

A Guide to the World's Most Maligned, Yet Misunderstood Creatures

Ross Piper

AN IMPRINT OF ABC-CLIO, LLC
Santa Barbara, California • Denver, Colorado • Oxford, England

Library of Congress Cataloging-in-Publication Data

Piper, Ross.
Pests : a guide to the world's most maligned, yet misunderstood creatures / Ross Piper.
p. cm.
Includes bibliographical references and index.
ISBN 978-0-313-38426-4 (alk. paper) — ISBN 978-0-313-38427-1 (ebook)
1. Pests. I. Title.
SB601.P49 2011
632'.6—dc22 2010041536

ISBN: 978-0-313-38426-4
EISBN: 978-0-313-38427-1

15 14 13 12 11 1 2 3 4 5

This book is also available on the World Wide Web as an eBook.
Visit www.abc-clio.com for details.

Greenwood
An Imprint of ABC-CLIO, LLC

ABC-CLIO, LLC
130 Cremona Drive, P.O. Box 1911
Santa Barbara, California 93116-1911

This book is printed on acid-free paper ∞
Manufactured in the United States of America

Contents

Introduction

Humans are but one animal species among millions, yet we are unique in the way that we have adapted the environment to our own needs. Our intelligence has enabled us to spread all around the globe and domesticate plants and animals for food, but these advances are not without their penalties. Humans, as do any other species, have a raft of parasites: animals that feed on us or in us, often causing harm. As humans have spread around the globe, travel between distant lands has become easier and the movement and spread of our parasites was similarly made easier. The plants we domesticated thousands of years ago were eaten by a myriad of herbivores; as we nurtured these plants, we increased the food supply of these herbivores, and they too spread wherever the crops were grown. Like us, the animals we domesticated had their own suite of parasites. In breeding these animals and spreading them around the world we gave their parasites a global meal ticket. There are also those animals that harm us and our domesticated animals in other ways (e.g., by injecting us with venom and causing illness and even death). In some parts of the world these venomous creatures can be a real problem to the extent where they are considered to be pests also.

The animals that annoy us, harm us, eat our crops, and torment our livestock are not inherently bad; they are merely doing what they have always done. Sadly, for them, they compete with us and we see them as a problem, so we do our level best to wipe them out. To us they are the pests and the most hated animals on the planet.

A huge range of animals, from nematodes to birds, are considered to be pests in various parts of the world. Some pests are limited to certain regions while others are more cosmopolitan in their distribution, but all of them are considered to be a nuisance in one way or another. To cover all of the animals that we view as pests in any amount of detail would require a small library, so the purpose of this book is to present a representative selection of these animals. Many books that deal with injurious animals are specific to either crop pests or pests of medical/veterinary importance, but in this book selections from both groups are included, which may aid in the understanding of pest science, the origins of pests, their impact on humanity, and how they help us to appreciate our far-reaching influence on the world around us.

The vast majority of animals are invertebrates, so it's no surprise that most pests are similarly lacking in backbone. The majority of animals featured in this book are insects and other arthropods, groups that include the most important pests of agriculture and human and animal health. Each entry in the book gives an overview of the pest in question and then looks in more detail at its biology, the damage it causes to warrant being classified a pest, and the measures that are employed to control the animal. This book is not a condemnation of these animals for the damage they cause; rather it attempts to provide a balanced view of how human activities have shaped the environment and are directly responsible for the pest problems we face today.

At the foot of each entry there is a further reading section that allows readers to find out more information about the species that interest them. In addition to book or journal resources there is also a huge amount of information about pests on the Web, but the interested reader should be mindful of the source of this information. What Web sites offer in terms of the quantity and accessibility of information is occasionally overshadowed by a lack of veracity. At the back of the book there is a list of Web sites produced by reputable organizations and institutions that have an obligation to supply the public and experts alike with accurate information on many of the pests presented here.

Any book providing an overview of animal pests would not be complete if it didn't look at some recurring themes that are important in understanding pests. The rest of the introduction is dedicated to covering some of these themes in greater detail, which will, I hope, add context to each of the pest vignettes in the main body of the book.

PESTICIDES

A pesticide is any substance that is used to control a pest, either by simply killing the creatures in question, deterring them from feeding, or preventing them from reproducing. Since the advent of synthetic organic chemistry the diversity of pesticide compounds has exploded and now there is a myriad of substances for the huge variety of pests against which we wage a never-ending war. We will take a look at some of the more important types of pesticide later, but before we do, let's briefly consider the history of pesticide use.

An Illinois farmer applies a low-insecticide bait to his crop. The insecticide is targeted against western corn rootworms, which would normally feed on, and lay eggs in, these soybeans. (Agricultural Research Service/USDA)

Since the dawn of civilization, humans have sought to control the species perceived to be pests. Thousands of years ago it would have been obvious to our ancestors that crops and livestock were eaten or plagued by other animals, thus reducing yields and endangering the very existence of these nascent towns and cities during the lean times of winter. The relationship between disease and other animals such as insects and rodents would have been much less clear as it wasn't until the germ theory came along that we became aware of the microbes that cause disease and how they are spread.

The emergence of civilization and the understanding that crops were at risk from various animals would have stimulated the inquiring minds of our ancestors, prompting the question: "How do we get rid of them?" This question prompted the development of pesticides. It is known that ancient farmers in the Near East applied substances to their fields, such as elemental sulfur, very likely as a means of promoting growth and improving yields, but such material may have coincidentally suppressed or eliminated certain pest species. Our knowledge of agriculture thousands of years ago is very fragmentary and it is entirely possible the early agriculturalists used a range of natural products to control the pests feeding on their crops. These people would have possessed a very thorough understanding of the wild plants that grew in their homeland and the various characteristics of each, possibly including the ability to kill or repel pests.

If we fast forward several thousand years we would see that pesticides remained largely primitive until quite recent times. With the advent of intensive agriculture it became clear that pests could have devastating, famine-inducing effects on crop yields. Throughout the medieval period and beyond farmers relied on toxic compounds based on arsenic, mercury, and lead to kill crop pests, although by the 17th century, compounds derived from tobacco and other plants were beginning to make an appearance. It is not until the 20th century that we encounter the golden age of pesticides. The 1940s saw the emergence of synthetic pesticides, which were organic molecules that stemmed directly from the great strides in chemical synthesis during the latter part of the 19th century and the early part of the 20th century. The scientific advances of the 19th and 20th centuries shed light on the nature of disease and a large number of parasitic species were added to the long list of animals that could be targeted with pesticides. Below, we'll take a brief look at some of the more important pesticides, beginning with the most widely used group, the modern insecticides and acaricides.

Insecticides and Acaricides

Insecticides are the most widely used pesticides because so many insect species feed on our crops and livestock and transmit disease to us and the animals we have domesticated. These same chemicals, when used to control mites and ticks, are known as acaricides. The most important insecticides / acaricides are briefly presented below.

Organochlorines

The most well-known insecticides are organochlorines, a class of synthetic, chlorine-containing compounds, many of which proved very effective at killing insects. Organochlorine pesticides include such well-known names as DDT, aldrin, dieldrin, and lindane. Chemically, these compounds are very stable neurotoxins, which are very resistant to degradation, characteristics that were once considered extremely admirable in a pesticide. They kill pests by interfering with the way in which tiny channels on the surface of neurons work, stimulating them to repeatedly produce nerve impulses and effectively disabling the nervous system of the animal in question.

When they first appeared on the scene, organochlorines were heralded as the nail in the coffin of pests. By the 1940s and 1950s science had significantly advanced our understanding of the role played by vector arthropods in transmitting diseases, so the organochlorines were seen as wonder chemicals, capable not only of preventing crop losses but also as a means of eradicating arthropod-borne diseases such as malaria. Throughout the 1950s the production and use of these chemicals increased enormously and by the beginning of the 1960s thousands of tons of organochlorines were being sprayed and dusted all over the world each year. However, the organochlorine revolution was not to last. It gradually became clear these lipophilic chemicals were very resistant to degradation. They accumulate in living things, with drastic consequences, the full extent of which is gradually becoming clear. The devastating environmental consequences of widespread organochlorine use (see Pesticides and the Environment) led to the ban of DDT in 1972. The Stockholm Convention on Persistent Organic Pollutants ratified a global ban on DDT and many other organochlorine compounds in agricultural applications, but many of these compounds are still used, controversially, to control arthropod vectors of disease, such as mosquitoes.

Organophosphates

When organochlorines fell out of favor, another class of compounds, organophosphates, was quickly adopted as the pesticide of choice. Examples of organophosphate pesticides include chlorpyrifos, dichlorvos, and phosmet. Like the organochlorines, the organophosphates are neurotoxins, but they have a distinct mode of action based on the inhibition of the enzyme that breaks down the neurotransmitter acetylcholine. With too much acetylcholine in the junction between nerve cells, nerve impulses are continually generated and nerve function is impaired to such an extent that a large enough dose causes death. Initially, organophosphates were seen as ideal substitutes for the very effective organochlorines. They were shown to be effective against many different types of pest and they degraded much faster than the organochlorines, overcoming the problem, it was thought, of environmental persistence. It has since become clear that organophosphates are far from safe. Their persistence is much less than organochlorines, but is sufficient for these chemicals to have significant, detrimental effects on nontarget organisms. Acute and chronic exposure to organophosphates is thought to cause disease and developmental defects in humans and wildlife. Organophosphates are still widely available despite the growing body of evidence demonstrating their detrimental effects on the health of humanity and the ecosystem as a whole. Organophosphates are used in a huge range of products, including those used in the home; however the tide of opinion will probably result in a global ban of organophosphates at some point in the not-too-distant future.

Carbamates

The first carbamate, carbaryl, was introduced in the 1950s. Carbamates have a mode of action similar to the organophosphates and are known to control a large range of insect pests. This broad spectrum of activity coupled with the carbamates' relatively low mammalian oral and dermal toxicity has seen them incorporated into many products, many of which are used in the home and garden. The chemical structure of carbamates makes them extremely toxic to the hymenoptera (bees, ants, and wasps) and they should be used in such a way as to protect honeybees and the myriad parasitic wasps that act as biological control agents of insect pests.

Plant-based Insecticides

Compounds derived from plants have been used to control crop pests for a very long time, but exactly how long is not known. Perhaps farmers of ancient civilizations noticed that certain wild plants were not really affected by herbivorous animals by virtue of the production of various compounds that deter or kill herbivores. The ability of tobacco derivatives to kill crop pests via alkaloids in the plant's tissues has been known for a long time. These alkaloids, of which the best-known is nicotine, are neurotoxins that block the transmission of electrical impulses through the nervous system. These alkaloids are very toxic to mammals as well as insects, so chemists used nicotine as a basis to create the nicotinoids and neonicotinoids, which are less toxic to mammals while retaining their potency against insects. The neonicotinoid imidacloprid is a relatively recent, yet very widely used insecticide and is often used to treat seed before it is sown; however, there are many experts who argue that the widespread use of neonicotinoids may be an important factor in the decline of honeybee populations around the world.

Another very widely used class of plant-derived insecticides is the pyrethrins, which are extracted from certain species of *Chrysanthemum,* notably *C. cinerariaefolium.* These compounds act in a similar way to the organochlorine compounds and they are also very soluble in lipids, but this is where the similarity between the two classes of insecticide ends. Pyrethrins are very unstable compounds and they are quickly degraded by exposure to oxygen, sunlight, and microbes. This very low environmental persistence is the reason why pyrethrins and their synthetic derivatives, the pyrethroids, are now the pesticides of choice in many applications in agriculture, in public health, and around the home. To improve the effectiveness of pyrethrins their structure was tweaked by chemists and the pyrethroids were born, examples of which include permethrin, cyfluthrin, cypermethrin, and deltamethrin. The main advantage of pyrethroids is their greater stability. They provide a more lasting effect than the pyrethrins, but still show considerably less environmental persistence than the organochlorines. Both the pyrethrins and pyrethroids are extremely toxic to aquatic life and hymenoptera.

The neem tree (*Azadirachta indica*), native to the Indian subcontinent, is the source of the insecticide, azadirachtin. Azadirachtin acts as a potent antifeedant and growth disruptor with considerable toxicity to insects. In contrast, its toxicity to mammals and other vertebrates is low, making it one of the safer insecticides. Also, being plant derived, azadirachtin has very low environmental persistence. Derivatives of the neem tree have probably been used for thousands of years in the Indian subcontinent in a range of applications and it is likely that early agriculturalists used the oil from the pressed seeds and leaves to help control crop pests.

Rotenone is extracted from the roots of various species of tropical and subtropical leguminous plants. Today it is produced commercially from extracts taken from the roots, leaves, and seeds of these plants. Like many other insecticides it is fat-soluble, enabling it to pass into the insect's body through the tiny gas-exchange tubes known as trachea. Rotenone is a potent insecticide that acts by interfering with cellular respiration in the mitochondria of the target animal's cells. In contrast to some of the other insecticides it is also rather toxic to vertebrates, especially fish, and because of this it is often used as piscicide to control fish that are considered to be pests for one reason or another. Even though it is toxic to vertebrates its environmental persistence is very low because it is rapidly broken down by sunlight. In addition to its potency as an insecticide and a piscicide, rotenone is also used to kill mites and ticks.

Antihelminthics

Helminth is a rather obsolete name for the huge group of animals that includes the nematodes and the platyhelminthes (flukes, tapeworms, etc.). The term *anithelminthics* typically refers to any chemical that is used to treat parasitic nematode and platyhelminth infections.

Avermectins and their synthetic derivatives, the ivermections, are commonly used as anthelmintic drugs and are typically given to livestock to kill gut parasites as well as parasitic insect larvae. Avermectins, like the majority of insecticides, are neurotoxins, but they have an inhibitory effect on the nervous system rather than a stimulatory effect and in high enough doses they kill the target animal. Avermectins are used routinely to treat livestock, pets, and occasionally humans and they are known to very effective at reducing the burden of intestinal parasites.

The benzimidazoles are a class of chemicals that have been used to eradicate parasitic nematodes and platyhelminthes from the bodies of humans and animals since the 1960s. These chemicals cause the death of the target worms by compromising the internal cell scaffold, which gives these chemicals a very broad spectrum of activity.

Piperazine has been used as an anthelmintic for around 50 years and it appears to rid the body of intestinal nematodes by acting as a neuroinhibitor. Parasitic nematodes exposed to sufficiently high doses of this compound become flaccid and lose their grip on the intestinal wall, eventually passing out of the anus of the host.

Rodenticides

As many rodent species around the world are considered to be pests, they have their very own pesticides. These rodenticides are typically anticoagulants—substances that act by inhibiting the clotting abilities of the blood, a crucial physiological phenomenon to preserve the integrity of the circulatory system and the life-sustaining functions it fulfills. These rodenticides are often used as baits—material the rodents in question will eat or gnaw, thus ingesting a dose of anticoagulant sufficient to cause lethal internal bleeding.

The Future of Pesticides

Pesticides are and will remain an important part of pest control simply because they are the cheapest means of controlling pests over large areas. Organic chemists will continue to design compounds that kill pests, all the time aiming to produce chemicals that are potent but with an acceptable level of environmental toxicity. The past has shown us that the full extent of a pesticide's impact on the environment may only be realized several years or decades after their introduction. The worrying fact is that we still don't fully understand how these chemicals can influence the behavior and physiology of other animals, including ourselves. Perhaps in 50 or 100 years' time pesticides will be considered obsolete and dangerous in light of other scientific and technological advances.

PESTICIDES AND THE ENVIRONMENT

Throughout the 1950s some scientists began to voice their concerns about the widespread use of synthetic insecticides, but the momentum generated

The thinning of brown pelicans' egg shells exemplifies the dangers of using DDT, which is now banned in many countries. (U.S. Fish & Wildlife Service / Steve Van Riper)

by dramatic results and corporate-sponsored research went some way to drowning out these fears. It wasn't until 1962 with the publication of *Silent Spring* by Rachel Carson that the concerns of many were presented in a way that was accessible to people other than scientists. In *Silent Spring,* the devastating ecological impact of organochlorines was exposed.

Organochlorines—A Persistent Problem

Organochlorines, such as DDT, are very soluble in lipids; therefore a small animal such as a caterpillar exposed to a sublethal dose of DDT will accumulate the compound in the fatty deposits of its body. When a small bird such as a sparrow eats 50 or 100 hundred such caterpillars it will accumulate the DDT in its fat tissue at a much higher concentration than was in the caterpillars it ate. When a top predator such as a raptor eats 50 or 100 such sparrows it receives an enormous dose of DDT. In birds especially the effects of DDT were unparalleled. DDT is not efficiently metabolized in animals and it builds up, interfering with many physiological processes, such as calcium metabolism—crucial in birds for the formation of the eggshell that protects the developing young. A bird heavily contaminated with DDT lays eggs with very thin shells that crack under the slightest pressure and the embryos within die. Organochlorine use caused significant declines in bird populations as well as other animals, effects that prompted the outright 1970s ban on agricultural use in the United States.

To this day, more than 30 years after the widespread use of organochlorines was banned, large mammals including humans are contaminated with high levels of these compounds. Breastfeeding mothers inadvertently feed their babies organochlorines as the compounds accumulate in breast tissue and its lipid-rich secretions. In the arctic, large mammals contain such high levels of organochlorines and other persistent organic pollutants that their washed-up bodies are sometimes classed as hazardous waste. The effects of these persistent organic pollutants on human health and ecosystem functioning are poorly understood, but they have been implicated as causative agents of some of the world's most important diseases, such as diabetes, cardiovascular disease, and cancer. Most worrying of all is that organochlorines are still being produced even though we know they accumulate and cause damage, the full extent of which is unknown, in all animals.

Avermectins and Ivermectins—Disaster for Dung Fauna

The avermectins and their synthetic derivatives, the ivermectins, are also of considerable environmental concern. These compounds are used to treat parasitic worm and insect infections in livestock, pets, and occasionally humans. In more affluent countries they are often used prophylactically to prevent the animal(s) in question from becoming infected in the first place. Initially thought to very safe, it is becoming increasingly clear that avermectins and their derivatives are far from innocuous in the environment. They are relatively stable compounds and often find their way into the environment via the feces of the treated animal. Livestock produce huge quantities of dung, which is a valuable resource for many animals, including countless invertebrates that depend on it for food. In turn these invertebrates are food for a huge range of vertebrates, including birds, terrestrial mammals, and bats. The avermectins and ivermectins are potent enough to kill the invertebrates that seek to take advantage of dung once it leaves an ungulate. Drastic reductions in this dung fauna has huge ramifications further up the food chain and in areas where these chemicals are routinely used there have been notable declines in birds and mammals.

Tributyltin—Marine Gender-bender

Another example of the devastating consequences of widespread pesticide use is the compound known as tributyltin (TBT), a substance that is used for many applications, including timber treatment and as an antifouling additive in ship paints to prevent the settling and growth of aquatic organisms. Over the years, significant quantities of TBT have found their way into the ecosystem and only in recent times have their physiological effects become apparent. Marine molluscs, especially gastropods, seem to be very sensitive to these compounds and at sublethal concentrations they can have very damaging effects. One of these is the strange condition known as imposex, where a female gastropod develops male sexual organs and vice versa, with obvious consequences for reproduction. TBT is very fat-soluble and relatively stable; therefore it is known to accumulate in the livers of large marine mammals, but it is still not known what effect this substance has on these animals.

Pesticides and Environmental Protection

Organochlorines, avermectins, and TBT show just how damaging our profligate use of synthetic pesticides has been for the environment as a whole. It's very likely that the known extent of the pesticide problem is only the tip of the iceberg. Honeybees, fundamental in the pollination of a huge number of crops, have been found to contain around 120 different pesticides, of which the neonicotinoids are considered to be among the most troublesome. How this complex chemical cocktail affects the biology of the bee is unknown, but this figure goes to show just how pervasive these compounds are in the environment. After 50 or so years of use synthetic pesticides are everywhere, from the food we eat to the furnishings in our homes. What are these chemicals doing to us? There is a growing body of evidence to suggest that persistent organic pollutants such as pesticides have a hand in causing many diseases, but a great deal more research is needed to define their true impact on us and the environment on which we ultimately depend. In the future, scientists with the gift of hindsight may look back at the mid- to late 20th century with astonishment at how we poisoned ourselves and the planet so spectacularly.

Ever since the agricultural revolution, crop yields have increased to feed an ever-growing population. As this rate of growth accelerates, so will the pressure on farmers to wring every last ounce of cereal, potato, or beef from their land. Many farmers see pesticides as a cost-effective means of controlling pests and an aid to improving productivity, but the stark realization is that this approach is hopelessly short-sighted. The environmental cost of pesticides and the evolution of resistance in the target organisms necessitate a complete reappraisal of the trajectory on which we now find ourselves. Do we go on poisoning ourselves and other organisms and face the long-term consequences, or do we use our intelligence to live our lives more in tune with nature?

PESTICIDE RESISTANCE

Nature's strength lies in its adaptability and it is this trait that has undermined the effectiveness of pesticides. If we consider the example of an insecticide being used to control the population of a beetle that is capable of ravaging a particular crop never previously exposed to such a chemical, we can imagine what might happen when we factor in genetic variability. Almost all of the beetles in the population will succumb to the insecticide.

However, there will be a tiny number of beetles with a chance genetic mutation that enables them to deal with the insecticide and break it down. Often this mutation is in a gene that codes for an enzyme involved in metabolism. In essence, these beetles are resistant to the insecticide. They will pass on the mutation that confers this resistance to their offspring and in a short space of time the beetle population will have recovered—made up entirely of individuals resistant to the insecticide.

The example above is just one way in which resistance to a pesticide can be conferred. In other forms of resistance, an individual in a population of a target species may possess multiple copies of a gene with instructions for producing an enzyme that breaks down the pesticide, rather than just one copy. With more of the enzyme the animal in question is better protected. In other cases, a target animal may possess mutations that result in behavioral changes as subtle as preferring places to rest that may protect the animal from the liberal application of pesticides.

Pesticide resistance is not limited to insects. Rodents, although not in the same league as insects when it comes to population growth, are still prolific breeders able to produce several generations per year. Over time, rodent populations, especially those of the brown rat, a serious problem in

The Colorado potato beetle has evolved resistance to many different types of insecticide. (iStockPhoto)

cities the world over, have become resistant to some of the anticoagulants used to control them. In many places around the world there are now rat populations with resistance to several types of anticoagulant.

Resistance is a huge problem confronting the widespread use of pesticides, which parallels the emergence of antibiotic resistance in bacteria. Insects and other animals such as rats are so abundant and their generation times are so often short that a mutation conferring resistance can be rapidly duplicated until an entire population of a given pest possesses it. The normal response in this situation is to switch to another pesticide with a different mode of action, a strategy that works in the short term until individuals with mutations conferring resistance to both compounds dominate the population and go on making a nuisance of themselves. In this way a pest can quickly develop multiple resistance, making it invulnerable to all the pesticides thrown at it. Another means of limiting the impact of pesticide resistance and extending the useful life of a particular product is by limiting their use to pesticide outbreak, rather than using them prophylactically.

PESTS AND ECOSYSTEMS

For the person seeking to protect crops or the person trying to safeguard human health, pests are nothing but a bad thing that need to be eliminated. This point of view is blind to our place in nature. Even in the 21st century with an understanding of the complexity of the natural world, we as a species still seek to control and dominate everything around us. From a purely biological perspective we are simply a dominant species taking over, but what sets us apart from the rest of the animal kingdom is our intelligence. Our inquiring minds have allowed us to recognize our place in nature: we are one cog in a complex machine and the damage we are doing to the environment will make the earth less able to support complex organisms such as ourselves.

Tsetse and the African Wilderness

The animals we call pests have been doing their thing for millions of years and they don't purposefully intend to harm us or eat our crops. The problem lies in our increasing insulation from the natural world and the way in which we have modified the environment. The existence of many species we know as pests is one reason some parts of the world retain areas

Intensive livestock rearing is still restrained in sub-Saharan Africa by numerous diseases, many of which are transmitted by insects. Gradual control of these diseases will result in a reduction in biodiversity as intensive livestock farming becomes more widespread. (FAO/18780/I. Balderi)

of wilderness. An example of how a pest can protect biodiversity is the humble tsetse of Africa. Much of low-lying fertile Africa is still free from intensive agriculture largely due to the impact of the tsetse rather than a magnanimous decision to preserve these treasure troves of biodiversity for subsequent generations. Multinational corporations have tried for some time to introduce productive, nonnative breeds of cattle into Africa to establish a cattle industry that is lucrative for everyone apart from the local inhabitants of these areas. However, these cattle have no natural immunity to the parasites and pathogens transmitted by biting flies like tsetse. No sooner are these cattle introduced than they succumb to the diseases transmitted by these flies.

Pest and Host Interactions—Hidden Complexity

Pest species inadvertently protecting natural habitat from development is one facet of the relationship these animals have with the wider environment. In this book, we briefly look at some of the nonsegmented worms that are internal parasites of humans and domesticated animals. These intimate relationships between host and worm have evolved over millions of years, yet the complexity of these interactions is poorly understood and the same goes for the relationships that exist between microbes and their hosts. The nematodes, trematodes, and cestodes have been part of vertebrate life for a vast length of time, as evidenced by some of their bewildering life cycles. Scientific research is beginning to tease apart the subtleties of these interactions. If we look at the way in which the immune system of humans deals with these parasites it appears that the two groups have evolved together for so long that there is an almost an element of mutual need. In Western, affluent societies where good health care has more or less eradicated many of these parasites, the incidence of immune system dysfunction, such as allergies, autoimmune disease, and cancer, is much higher than in developing countries where the parasite burden is far higher. Have we been overly hasty in trying to eradicate these parasites before we fully understand their inextricable and ancient links with us, their hosts? I'm not suggesting that people in developed countries should inoculate themselves with the eggs and larvae of parasites, but as with anything in nature the face value of a relationship belies its true complexity. These parasites undoubtedly cause disease in humans and other animals, but before we blindly try and eradicate them from the face of the earth let's try and figure out the intricacies of the relationship and what they mean for the immune system and disease.

Our Place in the Environment and Our Obligations

The examples above demonstrate the complexity of seemingly simple problems as well as making it painfully clear that humans are simply one animal among many, all of which share a planet and a common heritage. Some people may argue that the natural world is there for us to do as we please, but living in this way will eventually erode the very systems that keep us alive. Destruction of the environment is most often carried out by and on behalf of corporations, which may prioritize profits over careful consideration of environmental impact.

The degradation of our environment is accelerating, a result of a burgeoning human population. At the current rate of population growth there are around 70 million more people on the planet every year, all of whom need food, water, somewhere to live, and an infrastructure to supply all these things. This places a huge burden on rapidly dwindling natural resources. Massive leaps in science have provided us with ways of controlling disease and the vectors of disease, but at what cost? Without these natural limits on population growth, the number of *Homo sapiens* will grow at an ever-accelerating rate until our impact on the natural world is enough to make this planet inhospitable to human life.

When we stop to consider humans as just another, albeit intelligent, animal, we are faced with the brutal possibility that we, as a species, are a global pest. Our numbers increase unchecked and we wipe out many of the other species that share the planet with us. We consume natural resources and change the planet to suit our own ends with scarcely a thought for the delicate mechanisms that keep conditions on earth conducive to human survival.

If we are to avert a disaster of our own making in the future, we need a complete shift in thinking, beginning with recognition of our place in nature and commitment to living in harmony with the natural world. Our attitude toward pest animals perfectly demonstrates the growing gulf between humans and the natural world. The problems presented by pests would be less intense and the need to relentlessly pursue them with toxic chemicals would be much reduced if we could stem human population growth and produce food in a more sustainable manner.

MODERN AGRICULTURE

As the human population grows, the agricultural industry needs to increase production to supply the ever-growing demand for food. Since the agricultural revolution in the 18th century, agricultural productivity has increased, initially aided by improvements in techniques and then by advances in plant breeding, fertilizers, and pesticides. Today, many farmers believe that the most cost-effective way to feed the burgeoning human population is by dedicating huge areas of land to a cropping system known as monoculture. A monoculture is an area of land planted with a single crop. Monocultures make it easier for farmers to sow, manage, and harvest their crop. It is the most widely practiced agricultural system, but this extreme environmental homogeneity is in stark contrast to the heterogeneity of natural habitats where many species exist side by side.

The monoculture system is beset with problems, namely diseases and pests. An artificial environment that favors the growth of one species over all others is a perfect breeding ground for the organisms that feed on this

Monocultures are the basis of modern intensive agriculture and the crop yields from these systems are maintained with large inputs of fertilizers and pesticides. (Dreamstime)

crop and their numbers can swell enormously. Monocultures are defined by a reduction in biodiversity and the natural enemies and competitors of the organisms that destroy the crops are less abundant or even absent. With natural control severely limited, pests and diseases in these monocultures can abound. The situation is exacerbated by selective plant breeding that produces cultivars with high yields, but low resistance to these problematic organisms. In the early days of synthetic pesticides the problems presented by pests and diseases in monocultures were surmounted for a few years, but as resistance to these chemicals began to evolve, farmers were forced to use greater and greater quantities as well as new chemicals and chemical cocktails to achieve the same effect.

Integrated Crop Management

The environmental damage wrought by pesticides made it clear that such conventional cultivation with an over-reliance on chemical inputs was not sustainable in the long term. There has been something of a renaissance in using biodiversity and refined growing techniques to produce food in a sustainable manner.

This new approach is broadly termed integrated crop management (ICM). Although it's not really new, it does offer us a viable, environmentally sound means of growing food. ICM encompasses the following strategies:

- Crop rotations
- Appropriate cultivation techniques
- Careful choice of seed varieties
- Minimum reliance on artificial inputs such as fertilizers, pesticides, and fossil fuels
- Maintenance of the landscape
- Enhancement of biodiversity

Crop Rotations and Intercropping

Crop rotations are an important part of ICM as they increase the diversity of crop species, helping to prevent disease and limit the impact of pests. In a crop rotation a given area of land is rotated through different crops from year to year. One year the ground may be sowed with clover, a plant that uses symbiotic bacteria to convert nitrogen from the air into nitrates that enrich the soil for subsequent crops, such as corn. Crop rotations are extremely useful in preserving soil fertility and soil structure as well as minimizing erosion by ensuring adequate crop cover, good rooting depth, and reduction of soil compaction. In a crop rotation system, disease-resistant plant cultivars can minimize the need for inputs such as pesticides.

There is a huge variety of cultivation techniques available to farmers who want to explore ICM and the benefits it offers. Intercropping is another facet of sustainable farming, where two crops are cultivated together. The characteristics of each crop complement one another and growing both together is advantageous for both cultivation and pest and disease control. In some situations one of the cultivated plants may not be a crop as such. For example, leguminous vegetables such as peas can be planted alongside flowers such as marigolds. The strong odors produced by the marigolds make it very difficult for pests such as aphids to locate their host plant as they do so primarily by detecting and flying towards the odors produced by peas. The exact system of crop rotation and intercropping varies from region to region and also depends on the preferences of the farmer and his experiences.

Minimal Cultivation

Another key part of ICM is minimal cultivation, basically spending less time and energy preparing the ground for a crop. This may seem counterintuitive, as any gardener will know that the state of the ground is crucial in producing strong and healthy plants. However, the differences in yield between minimal cultivation and normal cultivation are outweighed by the benefits: reduced fuel usage, reduced soil erosion, and huge benefits for the soil-dwelling organisms, many of which help to keep the soil mixed and help to control pests such as earthworms and predatory beetles and spiders. In minimal cultivation strategies the only time when the farmer uses more conventional methods is in effective seedbed preparation, which enables the crop to become firmly established. Again, the type of minimal cultivation a farmer chooses depends heavily on the soil type, climate, topography, and individual preferences.

Reduced Chemical Inputs

Reducing inputs is instrumental in ICM and perhaps the biggest input in conventional farming is the use of fertilizers. Reducing the input of these chemicals is dependent on an understanding of individual crop requirements, particularly how much of the soil's nutrients a particular crop removes and therefore must be replaced. The amount of fertilizer already present in the soil as residues also needs to be assessed. To make all of these assessments, regular analysis of the soil is recommended, which provides the farmer with the information he needs to make a decision on how much or how little fertilizer he needs to apply.

The second largest input in conventional farming after fertilizers is pesticides, and although ICM doesn't advocate abandoning these chemicals, its success does hinge on their judicious usage. In situations where the use of pesticides is seen as unavoidable, a highly selective compound must be used carefully to limit the damage to nontarget organisms, many of which are crucial as predators of plant pests. Often, the numbers of a pest may not grow to a size where they cause economic damage to a crop—the so-called economic threshold. To assess pest populations the farmer can place traps in the crop to determine if pesticide application is appropriate.

Restoring the Balance of Nature

ICM also aims to reduce pesticide inputs by creating or restoring habitats that are conducive to the survival of natural enemies of plant pests. These natural enemies include organisms as diverse as birds and parasitic fungi. Minimal cultivation ensures that the microhabitats required by predatory animals such as beetles and spiders are not disturbed, leaving their populations intact. Both the adults and larvae of rove beetles and ground beetles are voracious predators of plant pests in agricultural environments, but they are very sensitive to disturbance and the effects of insecticides. Similarly, spiders require a heterogeneous habitat in which to hunt effectively and such habitats are encouraged in ICM.

The habitats surrounding a crop can be managed in a way that makes them attractive to animals that feed on or parasitize plant pests. For example, plants producing nectar-rich flowers can be encouraged in field boundaries as these are used as a food source by the myriad wasp species that prey on plant pests. Similarly, many species of plant often referred to simply as weeds produce seeds that attract birds, who then feed on any invertebrate pests they can find. Woodlands, hedgerows, and ponds in close proximity to crops should be encouraged and managed sympathetically for wildlife as they make the cultivated environment more heterogeneous, which in turn increases biodiversity and enhances the populations of natural enemies.

Integrated Pest Management as a Part of ICM

The cultivation techniques, reduced inputs, and habitat management of integrated crop management also form part of the strategy known as integrated pest management (IPM). Integrated pest management came into being in the 1960s, prompted by the large-scale failure of insecticides, specifically in cotton production, where more than 12 sprayings of insecticide per crop were used to control the devastating insect pests of this important plant.

For an IPM strategy to be successful requires a thorough understanding of crop fauna, both the crop pests and the natural enemies. This knowledge needs to include the basic ecology of the species in question and how they respond to their environment. Therefore, scientists will need to build an understanding of the reproductive capabilities of these animals, how they interact with other species, and how environmental variables such as weather, soil, and the availability of water, nutrients, and shelter will impact their numbers. Accumulating this information is time-consuming but also hugely instructive in developing ways of controlling pests without resorting to chemicals. An important part of IPM not already discussed above is the release of natural enemies such as wasps, flies, nematodes, fungi, bacteria, and viruses, all of which are known to attack the pest in question. Biological control of this type is known to work very well in closed environments such as glasshouses where the biological control agents are confined to a specific area. In the open landscape, some biological control agents work less well as they tend to disperse before they do their job. Other methods of control that can work well in IPM strategies include various types of trap, which can be enhanced by incorporating pheromone attractants, although these are only available for a small number of pests. These pheromones can also be used to interfere with the reproductive behavior of the pests in question.

A well-thought-out IPM strategy has significant economic and environmental benefits, but many farmers even today are reluctant to adopt this approach. IPM strategies can only be founded on in-depth biological research of a pest and its natural enemies and all too often the necessary information is lacking or fragmentary. Also, IPM is perceived to be complex, especially when compared to simply spraying a crop with pesticides. Often, it is mistakenly assumed that IPM is used in place of conventional pesticides, but this is not the case. IPM strategies do incorporate insecticides, but the insecticides are used much more judiciously.

ICM as a Sustainable Means of Growing Food

ICM is a whole-farm approach to growing crops because it looks at the wider environment and asks how nature can be harnessed to help produce food in a sustainable way. Many farmers are still skeptical about the benefits of ICM, but the figures speak for themselves. Generally, ICM is associated with a 5–15 percent reduction in yields, but as the farmer's experience grows, yields become more comparable to conventional cultivation systems. However, when we line this yield reduction up against the savings made in ICM systems and the benefits to the environment it becomes clear that conventional cultivation can't really compete. Integrated crop management reduces costs by around 20–30 percent, reduces pesticide and fertilizer usages by 30–70 percent and 16–25 percent, respectively, as well as preserving the overall quality of the end product—the crop. Perhaps the most important consideration of all for farmers who are increasingly well-versed in business is that ICM maintains or even slightly increases gross profit margins. ICM therefore provides us with a way of growing food that is less time-intensive, less land-intensive, and ultimately much better for the environment.

FURTHER READING

Carson, R. *Silent Spring.* Houghton Mifflin, Boston, MA, 1962.

Hamilton, D., and S. Crossley (eds.). *Pesticide Residues in Food and Drinking Water.* Wiley, London, 2004.

Hond, F. et al. *Pesticides: Problems, Improvements, Alternatives.* Blackwell Science, London, 2003.

Levine, M. J. *Pesticides: A Toxic Time Bomb in Our Midst.* Greenwood, Westport, CT, 2007.

Mason, J. *Sustainable Agriculture.* CSIRO Publishing, Collingwood, Australia, 2003.

Radcliffe, E. B., W. D. Hutchison, and R. E. Cancelado (eds.). *Integrated Pest Management: Concepts, Tactics, Strategies and Case Studies.* Cambridge University Press, Cambridge, NY, 2008.

Ware, G. W., and D. M. Whitacre. *The Pesticide Book.* Meister Publishing, Willoughby, OH, 2004.

Whalon, M. E., D. Mota-Sanchez, and R. M. Hollingworth (eds.). *Global Pesticide Resistance in Arthropods.* CAB International, Wallingford, United Kingdom, 2008.

Arachnids

Chiggers

Barely visible to the naked eye, chiggers, also known as red bugs and harvest mites, are larval mites belonging to a number of species in the family trombiculidae. The most important species are *Trombicula alfreddugesi, T. autumnalis, T. splendens,* and several members of the genus *Leptotrombidium* (see table). Like all mites, chiggers have a fascinating life cycle comprising several stages, the significance of which is still poorly understood. These mites are only problematic for part of this life cycle, but that doesn't really detract from the annoyance they are capable of causing.

Female chiggers deposit their eggs on the ground in soil or amongst leaf litter. After around six days the egg splits open to reveal an inactive stage, the deutovum. After another six days the deutovum develops into an active, six-legged larva, which is the chigger—the only stage in the life cycle of these mites that feeds on other animals. The larva (chigger) must locate a host and it does this in the same way as ticks, by waiting for a suitable animal to wander by so it can clamber aboard. Chiggers aren't fussy when it comes to hosts and they will quite happily feed on a wide variety of vertebrates, including amphibians, reptiles, birds, and mammals. Chigger feeding is a remarkable process. Unlike many ectoparasites they don't feed on blood. Instead, they pierce the skin and inject saliva into the underlying tissues, killing and digesting the host cells, turning them into a nutritious soup that can sucked up by the mite along with the fluids surrounding the host cells. Other components of the saliva act on the cells surrounding the damage, hardening them to form a tube that the chigger uses as a drinking straw to access more host soup, so to speak, until the chigger is fully engorged after three to five days. Replete with food, the chigger drops off, leaving the parasitic way of life behind, and it enters another inactive stage, the nymphochrysalis. Two more stages follow, an eight-legged nymph and a further resting stage, the imagochrysalis, which gives rise to the eight-legged adults. These adults, like the

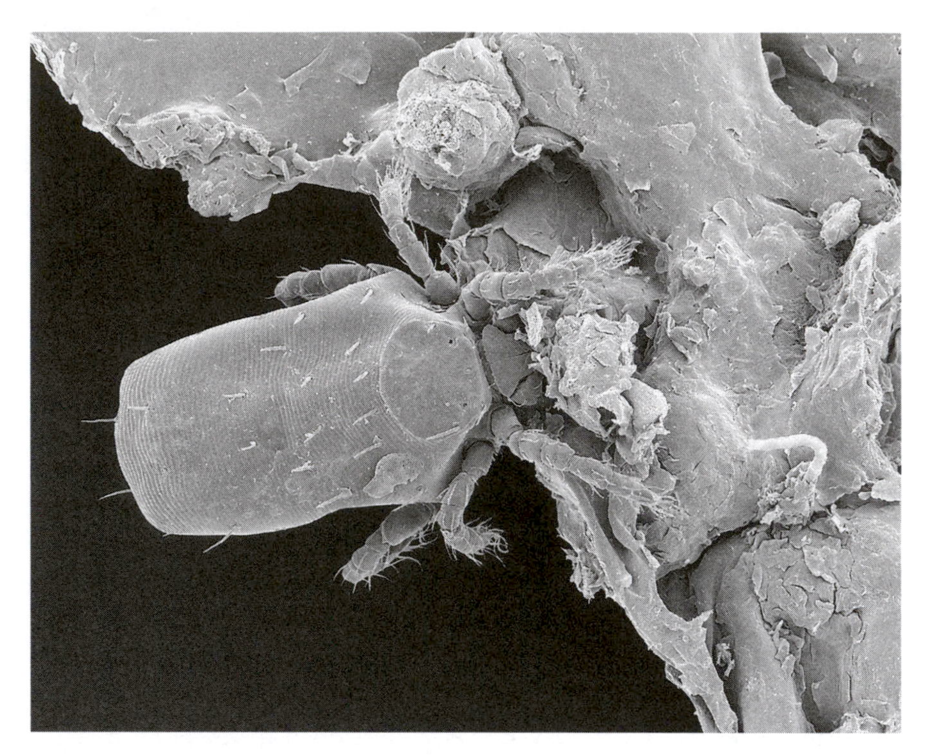

Chiggers are the larvae of certain mite species and they can cause severe irritation when they bite humans. (Dennis Kunkel Microscopy, Inc./Visuals Unlimited/Corbis)

larvae, are fearsome (albeit tiny) predators of various invertebrates, such as springtails, in the leaf litter and upper parts of the soil. Such a complex life cycle takes time to complete and depending on the species and the geographic location it can take between two months and a year. In temperate locations, there are one to three generations per year, while in the tropics there are continuous generations.

Chiggers do feed on humans, but we are accidental hosts rather than preferred hosts and the problems caused by these larval mites don't really extend past their nuisance value and the skin complaints they cause. However, in certain areas they are known to transmit disease to humans. Humans typically pick chiggers up from outdoors when the chiggers are waiting in vegetation for their preferred hosts to wander by. If these immature mites do find themselves on a human they make for areas where

The Important Chigger Species around the World

Species	Range	Natural hosts	Habitat	Period of larval activity
Trombicula alfreddugesi	Western hemisphere—Canada to Argentina and West Indies	Amphibians, birds, reptiles, and mammals	Typically edge habitats, i.e., woodland edges	Summer and early fall in temperate areas, year round in the tropics and subtropics
Trombicula autumnalis	Europe	Mammals, especially rabbits and rodents, ground-dwelling birds	Grassland and cultivated land	Late summer and early fall
Trombicula splendens	Eastern United States	Snakes and turtles, but also found on other vertebrates	Similar to *T. alfreddugesi,* but abundant in wet areas, e.g., swamps and bogs	Similar to *T. alfreddugesi*
Leptotrombidium spp.	Central, southern, and eastern Asia and Pacific islands	Small ground-dwelling rodents	Typically edge habitats and abandoned or poorly tended agricultural land	Late summer, fall

clothing fits snugly against the skin, such as the elastic parts of underwear and the waistbands of shorts and trousers. Because of their small size their mouthparts are only sufficient to pierce thin skin, such as the opening of a hair follicle. The saliva that chiggers inject when they feed causes an immune response noticeable as small areas of raised, inflamed, and very itchy skin. Even after the chigger has finished feeding or has been dislodged by scratching, the inflammation and itching persist, occasionally for several days. Scratching of these tiny wounds can lead to secondary bacterial infections, which in very rare cases may become serious, but more normally leads to small, slowly healing wounds. There are rare reports of the saliva of chiggers causing nervous system symptoms in dogs, including partial paralysis.

Aside from the irritation they cause, chigger species in the genus *Leptotrombidium* are vectors of tsutsugamushi, also known as scrub typhus, a potentially serious disease caused by the bacterium *Orientia tsutsugamushi.* Small outbreaks of this disease as well as regular cases from year to year have been reported for many decades from central, southern, and eastern Asia, but in recent years the number of reported cases has increased sharply. Between 2001 and 2005, 1,889 cases of this disease were reported in Japan, whereas the number of cases reported from Korea during the same period of time was 17,451. In both cases, the greatest number of cases occurs during the autumn months as a result of agricultural workers being in close contact with their crops during the harvest. Agricultural land (especially that which has been left fallow) supports large numbers of small rodents, the preferred hosts of *Leptotrombidium* species larvae.

Chiggers are very numerous creatures in various parts of the world and in the vast majority of cases, infestations, even large ones, do not cause any significant problems. Because of their limited impact beyond simply being a nuisance it often makes no economic sense to try and control them. The simplest means of avoiding chigger bites is to steer clear of their preferred habitats during periods of peak abundance—typically late summer and early fall, although this depends on latitude. If scrub typhus becomes more of a problem in the future, then control of the chiggers that transmit this disease may be necessary, but as we have learned with the attempts at controlling related parasites, such as the *Varroa* mite and ticks, the parasitic arachnids are exceptionally difficult to control. The most successful course of action in limiting the impact of these animals on human and animal health is increasing our understanding of their biology and preventing them from biting wherever possible.

FURTHER READING

Bang, H. E., M. J. Lee, and W. C. Lee. Comparative research on epidemiological aspects of tsutsugamushi disease (scrub typhus) between Korea and Japan. *Japanese Journal of Infectious Diseases* 61(2008): 148–50.

Mullen, G. R., and B. M. O'Connor. Mites (Acari). In *Medical and Veterinary Entomology* (G. R. Mullen and L. A. Durden, eds.). pp. 433–93. Academic Press, San Diego, CA, 2009.

House Dust Mites

Mites are everywhere, even in our homes, and often in very large numbers. Any home will support many species of mite and some dwellings in urban locations have been found to support 19 species of these little arachnids. Of all these microscopic lodgers there are three species of dust mite in the home that are of special importance and which account for 90 percent of the house dust mite fauna:

- *Dermatophagoides pteronyssius* (European house-dust mite)
- *D. farinae* (American house-dust mite)
- *Euroglyphus maynei*

The ancestors of these arachnids evolved at least 20 million years ago to take advantage of the food resources on offer in the nests of birds and mammals. Around 10,000 years ago humans began to live more settled lives and the door was literally wide open for these arthropods to take up permanent residence with us. To these microscopic arachnids, our homes, particularly our beds, are nothing more than enormous bird nests.

In the home they feed on all the detritus that constitutes the dust against which many people fight an obsessive war, although the mites derive much of their nutrition from the shed human skin cells that make up the bulk of this material. Every day, a person sheds 0.5–1 grams of dead skin cells and several thousand mites are able to survive for several months on just 0.25 grams of this material, which means that any home or place of work is a veritable banquet for these tiny animals. The mite's digestion of this material appears to be rather inefficient as they eat their own fecal pellets up to three times over to maximize the extraction of nutrients from their food.

The density of house dust mite populations vary according to the state of the living conditions in the homes where they are found, specifically the temperature and humidity. The favored relative humidity of these mites is at least 65–70 percent because atmospheric moisture is where they get their water. When the relative humidity falls to less than 50 percent the mites can only survive for 6–11 days; however, the protonymphs (an inactive, immature stage in the lifecycle) and the eggs can resist desiccation and are able to survive longer periods of adverse conditions. Each

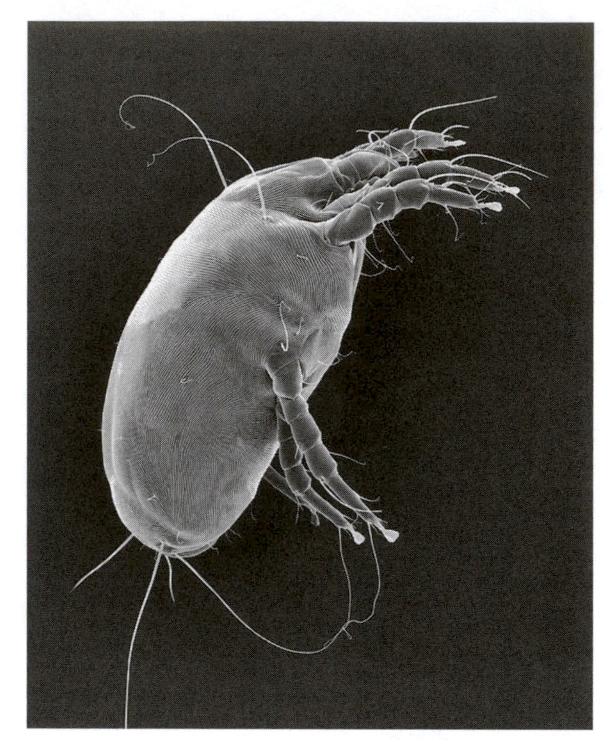

Millions of dust mites inhabit the home, feeding on dead human skin cells and fungi that are common in house dust. (Dennis Kunkel Microscopy, Inc./Visuals Unlimited, Inc.)

female house dust mite can lay 40–80 eggs and under perfect conditions (70% relative humidity and 23°C) development from egg to adult takes just one month.

Thanks to their small size, house dust mites are able to colonize new human dwellings very easily. They can be inadvertently transported on our clothes, in furniture, and in the fur of our pets. Once they've reached a new building their populations increase over a 10-year period as dust steadily accumulates. Eventually, the populations of these mites can reach staggering proportions. In heavily infested homes, a gram of dust from a mattress can contain 5,000 house-dust mites, the thought of which is enough to make even a bug-lover itch. The large numbers of dust mites a house can support is the reason why they can be a problem. Over time, bits of mite cuticle, dead mites, and fecal pellets, as well as the powerful

Dust Mite Species Commonly Encountered in Human Dwellings

Species	Geographic distribution
Dermatophagoides evansi	North America, Europe
D. farinae	Essentially worldwide (more common in North America than Europe)
D. halterophilus	Singapore, Spain, tropical regions
D. microceras	Europe
D. neotropicalis	Tropical regions
D. pteronyssinus	Essentially worldwide (more common in Europe than North America)
D. siboney	Cuba
Euroglyphus maynei	Essentially worldwide
Hirstia domicola	Present throughout studied regions
Malayoglyphus carmelitus	Israel, Spain
M. intermedius	Present throughout studied regions
Pyroglyphus africanus	South America
Sturnophagoides brasiliensis	Brazil, France, Singapore

enzymes they contain, all start to build up, ending up on the floor, on clothes, on bedding and furniture, and in the air. The fecal pellets of these arachnids are so small that if they are disturbed in an unventilated room during a frantic bout of cleaning, they remain airborne for up to 20 minutes, perfectly positioned to be inhaled. These bits and pieces of mite, particularly the enzymes contained within the fecal pellets, are very antigenic, which means they can elicit a strong immune reaction in sensitive individuals. One of the mite gut enzymes that finds its way into the fecal pellets of these arachnids is very invasive in the human body. This enzyme has been detected in the amniotic fluid of pregnant women and even in the blood of their unborn babies following its inadvertent inhalation by the mother. The presence of dust mites is also associated with a number of bacterial and fungal species that are able to thrive on the skin component of dust and the copious feces produced by the dust mites. Some of these microorganisms are themselves a public health concern as they have been implicated in various human diseases.

The propensity of these mite particles to trigger a powerful immune response is a direct result of two things—the indoor environment and our increasingly obsessive attitude toward hygiene. In recent decades humans have become obsessed with hygiene, which is no bad thing as far as reduction in the incidence of potentially life-threatening diseases. However, the downside to this obsession with hygiene is that our sophisticated mammalian immune system does not encounter challenges of the severity or frequency for which it was designed; therefore it becomes hypersensitive and reacts disproportionately to challenges that are inconsequential to our well-being. In many people, this disproportionate response manifests as asthma and related disorders such as allergic eczema, allergic rhinitis, and conjunctivitis. It has been found that house dust mite allergens act as a trigger for asthmatic attacks in 85 percent of people with this potentially lethal condition. On one hand, we have lessened our ability to effectively deal with challenges to our immune system and on the other we have provided organisms like the house dust mites with almost perfect living conditions where the humidity and temperature are more or less constant and food is plentiful.

The almost ubiquitous presence of house dust mites in the built environment around the world and their ability to trigger disease has made them the subjects of intensive research aimed at understanding the intricacies of their association with allergy and the ways in which their populations can be controlled. It is worth mentioning that eradicating these arachnids from dwellings is effectively impossible as viable populations can survive in the smallest recesses on tiny quantities of food. The best we can ever hope to achieve is control of house dust mite populations by making the built environment less attractive to these animals. Furnishings and fittings can be used that offer these mites fewer places to hide and provide fewer places for the accumulation of dust. In addition, improving ventilation in dwellings and reducing temperatures by a degree or two can make conditions less conducive to dust mite survival. Certain pesticides based on benzyl benzoate are also used in an attempt to control dust mite populations, but it is not clear how effective these products are or how their potential accumulation could affect the health of humans and pets inhabiting dwellings where they are routinely applied.

FURTHER READING

Arlian, L. Biology and ecology of house dust mite. *Dermatophagoides* spp. and *Euroglyphus* spp. *Immunol. and Aller. Clin of N. America* 9(1989): 339–56.

Walter, D. E. *Mites: Ecology, Evolution and Behavior.* CABI Publishing, Wallingford, United Kingdom, 1999.

Scabies Mite

The mites, *Sarcoptes scabiei,* that cause the disease known as scabies are microscopic creatures, which is somewhat of a blessing given their exceedingly unsavory appearance. No parasitic mite can be considered to be attractive, but the scabies mite really drew the short straw when it came to looks. They look like bumpy little sacks sprouting a few long bristles, trundling around on stumpy legs. The adult females are a little less than half a millimeter long, whereas the adult males are even smaller.

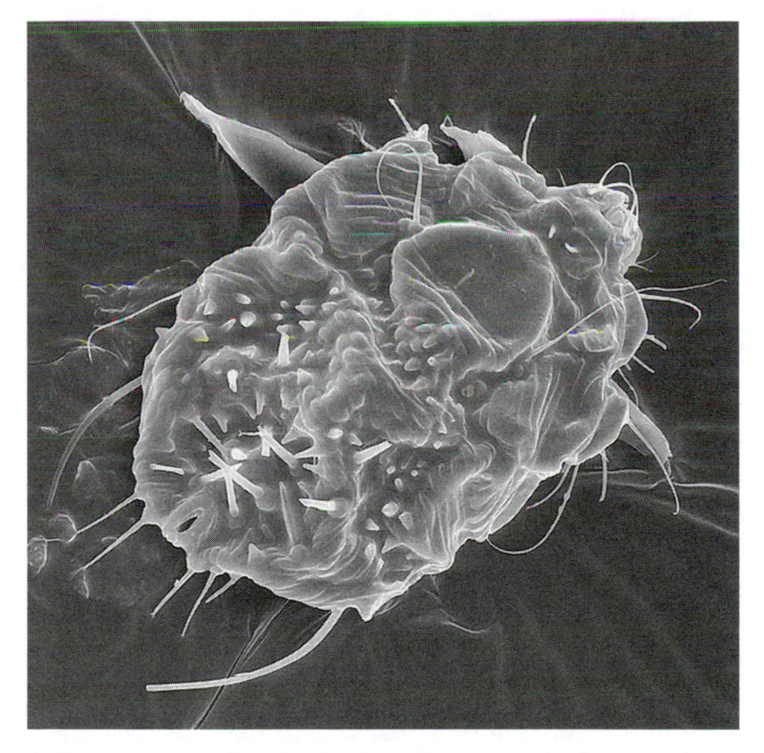

Scabies mites tunnel under the skin of humans and other animals causing intense irritation. (Visuals Unlimited/Corbis)

From a zoological point of view, these mites are obligate parasites of a wide variety of domestic animals and each host has its own variety of *S. scabiei.* For example, the variety encountered on humans is *S. scabiei* var. *hominis,* while the type found on horses is *S. scabiei* var. *equi.* These varieties are morphologically indistinguishable from one another and they certainly have no problem parasitizing domesticated animals other than the species they are typically found on, so it remains to be seen if they are genuinely varieties or simply all members of the same species in the very early stages of speciation. The evolutionary origins of these mites is very poorly known, but must be very interesting as they may be an ancient parasite of humans that has taken up residence on all the animals our species has domesticated in the last few thousand years. Alternatively, they may be a recent addition to the roll call of human parasites after they moved onto us from one of our domesticated menagerie. Regardless of the details of their evolutionary origins these mites have proved to be very successful parasites, able to survive and thrive on a range of hosts.

Mites often have fascinating life cycles and the scabies mite is no exception. Gravid females wander on the skin to find a suitable place to excavate a burrow. Despite their tiny size and stubby legs, they can move relatively rapidly—about 2.5 centimeters per minute. Once she's found a desirable spot, the female uses the long bristles on her hind legs to tilt her body so that she's almost vertical and then, like a miniature drill-bit, she uses her forelegs, mouthparts, and a skin-dissolving secretion to penetrate the surface of the skin and excavate a burrow in the upper parts of the epidermis. Each day, the female lengthens the burrow by about 0.5 millimeters, feeding on the skin epidermal liquid and laying eggs as she goes until she has excavated a winding tunnel around 1 centimeter in length. The eggs hatch and the larvae remain in their mother's burrow for about a day before crawling out onto the surface of the skin, where they excavate small burrows, known as molting pouches, in which they develop into nymphs and then adults. The time taken for an egg to develop into an adult is around 10 days for males and 14 days for the larger females. It is thought that less than 1 percent of the eggs laid will develop through to adulthood. The adult males leave their molting pouches and go about looking for mature females, who sit tight in their pouches. Mating takes place in the female's molting pouch and with her eggs fertilized she crawls out onto the surface of the skin to look for a suitable place to excavate a permanent burrow in which she will deposit her eggs to complete the life cycle.

Scabies is the disease caused by the burrowing activities of these mites and it is a problem for both humans and domesticated animals. In humans, the areas of the body typically affected are those that offer the mites sheltered, warm, and relatively humid conditions, such as the skin between the fingers, the skin between the shoulder blades and beneath the breasts, the skin around the genitals, and the creased skin of the elbows and knees. The main problem with these mites is the irritation caused by the burrowing behavior, which is actually an immune reaction to the fragments of mite cuticle, mite feces, and eggs in the burrows. In sensitive individuals, this immune reaction can be very severe, whereas in those people with a depressed immune system, such as HIV patients, the mites elicit no immune reaction whatsoever, allowing huge populations of these arachnids to build up.

Scabies is a global problem and the prevalence of this disease appears to be increasing, probably thanks to burgeoning urban populations and the poor, crowded conditions in which many city-dwellers have to live, especially in the developing world. The mites that cause scabies have no respect for age, gender, ethnicity, or social class, and the disease affects all types of people, from peasant farmers in Southeast Asia to wealthy bankers in North America. It has been estimated that at any one time 300 million people around the world have scabies and for reasons that are poorly understood the disease appears to follow epidemic cycles, ranging from 20–30 years between peak levels of infection, possibly due to changing levels of immunity in the human population.

As a disease, scabies is highly contagious. The mites pass from person to person via direct skin contact. Even when the mites are dislodged, they can survive on bed linen, clothes, and other fabrics for 24–36 hours at room temperature with normal humidity (21°C and 40–80% relative humidity)—even longer at lower temperatures with high humidity. In view of how easily these mites can be transferred from person to person it's no surprise that one case of scabies in a population quickly becomes many. In some Australian indigenous communities, 20–65 percent of people are infected with scabies mites and in other parts of the world the incidence can easily reach 90–100 percent during severe outbreaks in very crowded areas.

The itching caused by these mites and the lesions that form on the skin may not seem like a big deal, but various complications can develop as a direct result of scabies. Bacteria, especially group A streptococci and *Staphylococcus aureus,* can invade the lesions caused by the mites and the small wounds created by scratching, causing secondary infections. These infections can be easily treated with antibiotics, but the immune response to long-term infections can eventually lead to kidney and heart disease, both of which occur commonly in populations where mite infestations are commonplace. With all these generally unseen consequences it is extremely difficult to estimate the full impact of scabies on humans around the world, but we can be certain that scabies is a real and growing problem.

S. scabiei also affects a huge variety of domesticated and wild animals, where it causes the disease known as sarcoptic mange. Canines and cervids (deer and their relatives) are most commonly and seriously afflicted, whereas cats and guinea pigs are the only domesticated animals not to be troubled by this parasite. Livestock are often maintained at a high density, so mange can be a serious problem for the agricultural industry. Mange is considered to be a major problem in pig farming as growing pigs put on less weight when they are infested with these mites. Their appetites fall in response to the immune reactions and complications arising from secondary bacterial infections, which contributes to significant financial losses in this industry. Dogs can also be severely affected by mange, particularly those that have been abandoned, and it is not unusual to see animals in this situation that have lost most of their fur, their skin covered in thick crusts due to the mite infestations and secondary bacterial infections.

Scabies mites are very difficult to control. A number of commercial acaricides are available that kill the mites on the host, but if the root cause of the problem is not addressed, namely overcrowding and poor living conditions, both of which apply to human and domesticated animal infestations, there will be a continual cycle of treatment and reinfestation. Another barrier to the effective control of this parasite is the emergence of acaricide resistance. Mites are known for their ability to rapidly evolve resistance to acaricides and the causative organism of scabies is no exception. New acaricides are being investigated that show promise in the control of these mites, but the use of these chemicals must be accompanied by an understanding of the parasite's life cycle and improved living conditions for domesticated animals, as well as people, all of which contribute to eliminating the scabies mite from vulnerable communities.

FURTHER READING

Arlian, L. G. Biology, host relations, and epidemiology of *Sarcoptes scabiei*. *Annual Review of Entomology* 34(1989): 139–59.

Chosidow, O. Scabies. *New England Journal of Medicine* 354(2007): 1718–27.

Walton, S. F., and B. J. Currie. Problems in diagnosing scabies, a global disease in human and animal populations. *Clinical Microbiology Reviews* 20(2) (2007): 268–79.

Scorpions

These arachnids are synonymous with danger as every single species is venomous and is capable of penetrating human skin with their sting. With this said, of the 1,400 or so known species of scorpion, only around 25 are known to be dangerous and capable of causing human death. The vast majority of scorpion envenomations are due to a handful of these species.

The scorpions are an ancient, yet extremely successful group of invertebrates that retain many of the features that enabled their distant ancestors to forsake the marine environment and conquer the land more than 400 million years ago. During this transition and in the eons that followed, scorpions evolved a host of adaptations to enable them to thrive on the land, often in habitats that are inhospitable to many other animals. Like

Scorpions, like this *Buthus occitanus* use their venom with great effect to kill their prey and defend themselves. Only a small number of species are dangerous to humans. (Courtesy of Ross Piper)

all arthropods, they have a tough exoskeleton that prevents excessive water loss; theirs appears to be particularly watertight, enabling these creatures to conserve water more effectively than most other arthropods. This waterproof armor coupled with their proclivity for carrying out most of their activities during the night means that many scorpions species are arid zone specialists, able to live quite happily in habitats where water is in extremely short supply.

Not only are these animals able to withstand conditions that would wither many other animals, but they are also excellent parents. The females nurture their eggs until the young, miniature adults hatch and crawl on to her back so that she can carry on with her normal activities while still protecting her brood. It's adaptations like these that make scorpions so successful. In some desert environments, scorpions can reach huge population densities. In some locations, scorpions at their peak abundance can make up more than 85 percent of the total predatory arthropod biomass, with a density of 1,000–5,000 individuals per hectare. Just how they can exist at such high densities in seemingly unproductive habitats is something of a mystery, but it is known they convert prey into arachnid biomass very efficiently. Their metabolism is very slow and many species can survive for a year a more without food, but when they do eat they gain as much as one-third of their body weight from one meal thanks to the way that much of their digestion takes place externally. Their slow metabolism, low-energy hunting technique (ambush), and willingness to take a range of prey means these animals can thrive in very marginal habitats.

The single reason why these animals can be considered to be of public health concern is the venom they use to subdue their prey. The venom is produced in glands in the tail and is injected into the victim via the sharp sting at the tip of the tail. Their accuracy with this sting is remarkable even though their eyesight is relatively poor, but this is no surprise when we remember these animals depend on their sting to subdue their prey once it's been grasped by the pincers. The venom itself is a complex mixture of compounds that have a number of effects, including inhibition or modulation of the way in which electrical impulses are transmitted through the nervous system, thereby causing paralysis of the muscles. Scorpion venom is very different from snake venom because its evolutionary origins are probably the secretory products of anal glands belonging to the ancient ancestors of these arachnids, whereas snake venom can be thought of as highly modified saliva.

The Most Dangerous Scorpion Genera, the Potency of Their Venom, and Where They are Found

Genus	Toxicity of venom (LD/50–mg/kg)	Distribution
Leiurus	0.25	North Africa and Middle East
Buthus	0.9	Mediterranean and parts of the Middle East
Parabuthus	4.25–100	Western and Southern Africa
Hottentotta	1.1–7.9	Southern Africa to Southeast Asia
Mesobuthus	1.45	Throughout Asia
Tityus	0.43	Central and South America, Caribbean
Androctonus	0.08–0.5 (*A. crassicauda*) 0.32 (*A. mauritanicus*)	North Africa to Southeast Asia
Centruroides	0.26–1.12	Southern United States, Central America, Caribbean
Odontobuthus	0.19	Iran

LD/50 is explained in greater detail in the text below. See also the Snakes and Spiders entries

In countries such as Brazil, Mexico, Tunisia, and Morocco, there are several thousand cases of scorpion envenomation every year, often involving curious children whose hands or feet force the scorpion into a situation where it has to defend itself. In Brazil alone during a three-year period there were 6,000 reported scorpion envenomations, 100 of which ended in death. In the United States during 2006, more than 16,000 scorpion stings were recorded and in a single Moroccan province (El Kelaa des Sraghna) almost 12,000 stings were reported in a five-year period (2001–2005). The complete extent of scorpion envenomations is undoubtedly far in excess of the figures quoted above as the majority of victims are poor, rural dwellers with limited access to modern medical facilities and in many cases the symptoms of the scorpion sting may be considered too mild to seek medical attention. The symptoms caused by a sting vary both between and within scorpion species. The symptoms also vary according to the site of the sting and the age and health of the victim (for example, children are much more likely to die from a scorpion sting than healthy adults because of their lesser body mass). The typical effects of envenomation from one of the dangerous scorpions range from flu-like symptoms to death within one hour. Symptoms develop rapidly if the venom is introduced into the body in or near a blood vessel.

The toxicity of venoms is quantified with the LD50 (lethal dose) test, which indicates how much venom it takes to kill 50 percent of the test animals, typically mice. The lower the LD50 value, the more toxic the venom. Also, it's worth remembering that humans are much more susceptible than mice to many venoms, including those produced by scorpions, so the LD50 values for humans may be much lower than those quoted in the supporting table. You can see that the species with the most potent venom are those in the genus *Androctonus,* so these can be considered to be potentially the most dangerous scorpions. Venom potency is not the full story, however, as the chances of humans coming into contact with the dangerous species must also be considered.

As with all venomous animals, scorpions don't set out to purposefully harm humans. The venom they produce is primarily for the capture of prey. Because venom is biologically expensive to produce, scorpions are very judicious with its usage. Nonprey species are only stung when the scorpion feels the need to defend itself. The propensity of these arachnids for feeding on invertebrates such as ground-dwelling insects means they are often drawn to human settlements where easy insect prey and places to hide abound. It is in these situations where they are more likely to come into contact with humans, meetings that occasionally end in envenomations. Another feature of the biology of at least one scorpion species that contributes to them being a problem in and around human settlements is their ability to reproduce without the need for males. This phenomenon is known as parthenogenesis and among the dangerous scorpions it is known to occur in *Tityus serrulatus.* The ability of this species to reproduce asexually means that it only takes a single female to form a colony in any given area, and because its reproductive potential is effectively twice that of species that reproduce sexually (i.e., all the individuals in a *T. serrulatus* population can reproduce), it can out-compete other, often less dangerous scorpion species.

From *T. serrulatus* and many other scorpion species there has been an increase in the incidence of envenomations in recent decades, but this increase is a result of the burgeoning human population and the establishment and growth of settlements in areas of previously untouched habitat. A perfect example of human expansion into scorpion territory is Brasilia, which was constructed in the 1950s to open up the interior of Brazil. If the presence of scorpions cannot be tolerated, the simplest way to limit their numbers is to deny them hiding places, such as crevices in and around buildings.

To summarize, there are a few scorpion species that are potentially dangerous, but treated with respect these animals can actually be a positive presence in and around settlements, where they play an important part in regulating the populations of insects that spread disease and that damage crops and stored foods.

FURTHER READING

Polis, G. A. *The Biology of Scorpions.* Stanford University Press, Palo Alto, CA, 1990.

Spiders

Apart from a single, anomalous species, all of the 35,000 species of spider are carnivores that, for the most part, prey on other arthropods, typically insects. To aid them in their predatory ways, all spiders, except those species in the family uloboridae, are able to produce potent venom to subdue and digest their prey.

Spiders are a fascinating group of animals, but there any many people who are terrified enough by these arachnids to develop a phobia. Although some spiders may look creepy and spend most of their time in the shadows there are actually very few species that can harm a human with their venom. In most cases, the spider's fangs are simply unable to penetrate human skin and even if they could it is unlikely their venom would cause anything more than localized pain and swelling. Regardless of these facts, spiders generally receive a very bad rap in the media due to the small number of species that are able to penetrate human skin with their fangs and produce venom that is potentially dangerous to us. Compared to more mundane causes of death, fatal spider bites are very rare indeed; however, we will look at the impact of these invertebrates on humans in this section. Before we do so, let's try and keep things in perspective by remembering that the benefit of spiders to mankind vastly outweighs whatever injury they cause us. Spiders are fundamentally important parts of terrestrial ecosystems and their predatory ways are crucial for the regulation

A Brazilian wandering spider shows off its fangs by using a threat posture. (iStockPhoto)

of the populations of other arthropods, including the myriad insects that impact our lives in far more serious ways.

Spider venom is a complex concoction of compounds. Its primary function is swift incapacitation of the prey and certain enzymes within the mixture are able to initiate the process of digestion. Spiders are unable to swallow lumps of food via their tiny mouths, which also have a fine filtering system; therefore all digestion takes place outside the body. Some species are able to mash their prey up to hasten the digestion process, but others simply leave the prey's body more or less intact and digest its insides using the body of the victim like a macabre vessel. In both cases, digestion is primarily achieved by the regurgitation of digestive juices from the arachnid's digestive tract. The only way into the spider for the resultant prey soup is through its very tiny mouth.

The lifestyle of a spider dictates the toxicity of its venom. Generally, in those species that use webs for hunting, the web itself helps to subdue the prey, so the venom these spiders produce need not be superpotent. On the other hand are those spiders that don't use webs to ensnare their prey. To make sure their quarry doesn't escape and hide once it has been bitten, these spiders produce very potent venom. There are exceptions to this generalization, such as the widow spiders that build webs and are also known for producing potent venom. Of all the spiders currently known to science, only about 500 species can cause a bite that can be described as painful, and of these only 20–30 can be considered to be genuinely dangerous to humans. The spider genera generally considered to be the most dangerous are *Latrodectus* (widow spiders), *Atrax* and *Hadronyche* (funnel-web spiders), *Phoneutria* (wandering or banana spiders), and *Loxosceles* (recluse spiders). Representatives of other genera can cause painful bites; these include *Tegeneria* (house spiders and relatives), *Sicarius* (six-eyed spiders), *Cheiracanthium* (sac-spiders), *Lycosa* (wolf-spiders), *Steatoda* (false-widow spiders), *Argyroneta* (water spiders), and *Missulena* (mouse spiders).

In common with snakes, spider venom can be neurotoxic or necrotizing. Neurotoxic venoms impair the correct functioning of the nervous system, while necrotizing venoms cause the breakdown of tissue. The most infamous venomous spiders are the widows, of which there are several species. The black widow spider, *L. mactans,* is the archetypal venomous animal, with its distinctive markings and propensity for loitering in dark places. The bite of this species is not particularly painful and in most cases it probably goes unnoticed. The first real pain is felt about 10–60 minutes after the bite in the vicinity of lymph nodes and from here the sensation of pain spreads to the muscles. The venom of this species is a potent neurotoxin that acts on the junctions between nerve fibers and muscles (neuromuscular junctions) as well as junctions (synapses) between nerve cells in the central nervous system. Certain compounds in the venom block the transmission of the electrical nerve impulse, essentially resulting in muscle paralysis. A bite from a widow spider can be dangerous if the muscles controlling breathing are affected. If the venom doesn't reach these muscles the patient is in much less danger. Without any treatment, the acute symptoms of a black widow bite will last for around five days and in the vast majority of cases the victim will make a full, albeit slow, recovery over several weeks. About 60 years ago, before antivenins were available to treat black widow bites, envenomation from this species caused death in about 5 percent of cases in the United States. This figure is now less than 1 percent. Interestingly, some domestic animals, such as horses, cows, and sheep are more sensitive to the venom from a black widow than humans, while for rats, rabbits, dogs, and goats, the opposite is true.

The funnel web spiders of the genus *Atrax* and *Hadronyche* are not far behind the widow spiders in the notoriety stakes. Denizens of Australasia, these are among the very few more primitive spiders (mygalomorphs) that produce potentially dangerous venom. Like the venom of the black widow, *Atrax* and *Hadronyche* venom contains neurotoxic compounds. It is interesting to note that only bites from male funnel web spiders cause potentially dangerous symptoms. This is because the male's venom contains a substance known as *robustoxin.* More interesting still is the fact this compound is only toxic to primates (virtually all domesticated animals are immune to the venom of these spiders). Why this should be is not clear, but it may simply be a quirk of evolution. Envenomation by a male funnel web spider causes severe pain, muscle cramps, temporary blindness, shivering—and, most seriousparalysis of the muscles in the thorax. Although the bite of the Australian funnel webs is undoubtedly cause for concern it should be noted that only around 12 fatal cases have ever been recorded. As Australia is home to such a large number of venomous animals, the medical authorities there have a well-honed system for dealing with envenomations and treating the patient in the appropriate way.

The wandering spiders (genus *Phoneutria*) of the neotropics are also sometimes erroneously known as banana spiders for their very occasional and accidental association with shipments of these fruits. Typically, the large spiders encountered in banana shipments are not *Phoneutria* species at all, but harmless look-alikes. *Phoneutria* species are perhaps the most dangerous spiders on the planet because they are very aggressive and they are capable of injecting relatively large amounts of exceedingly potent venom. The vast majority of spiders will only bite defensively, but it seems the *Phoneutria* wandering spiders will go out of their way to bite, especially if they feel threatened. To these arachnids, attack is the best form of defense. There are reports of these spiders scampering up the handle of a broom to bite the person trying to shoo it away and captive specimens throwing themselves at the glass of their terrarium when a person enters the room.

Unlike the bite of the black widow, the bite of a wandering spider is immediately painful and the victim can go into shock. The complex cocktail of compounds in *Phoneutria* venom causes rapid heartbeat, high blood pressure, profuse sweating, shivering, salivation, nausea, vomiting, vertigo, visual disturbances, and priapism (especially in boys less than 10 years old). If death occurs, it is usually within 2–12 hours. A bite from a wandering spider should be treated as a medical emergency as the patient will need antivenin as soon as possible. Although wandering spiders in the

The Geographic Distribution, Venom Yield, and Venom Potency of the Spider Genera Considered to be Most Dangerous to Humans

Genus	Common name	Body length (mm)*	Distribution	Venom yield (mg)	Venom potency – LD50 (mg/kg)
Atrax	Sydney funnel web spider	24–32	Australia	0.25–2	0.16
Hadronyche	Funnel web spiders	40–50	Australasia	No data, but more than above	Equivalent to or even greater than that of *Atrax*
Latrodectus	Widow spiders	8–15	Essentially world-wide except high and low latitudes	0.02–0.03	0.002
Phoneutria	Wandering spiders, armed wander-ing spiders, armed banana spiders	30–50	South and Central America	0.296–1.079 (dependent on sex and age)	0.00061–0.00157
Loxosceles	Recluse spiders	6–10	The Americas, sub-Saharan Africa, and southern Europe	0.13–0.27	0.48–1.45

*Size range for females are stated. Male spiders are always smaller than the females and produce less venom.
See Scorpions and Snakes entries for explantation of LD/50

genus *Phoneutria* are undoubtedly dangerous, data on the exact number of confirmed human fatalities is hard to come by. The growth of the human population in Brazil means that people are coming into contact with these spiders more frequently and each year there are thought to be 600–800 *Phoneutria* bites in the vicinity of São Paulo alone. *Phoneutria* spiders are nocturnal and they will often seek refuge in and around homes to wait out the day. It is in these situations that they are most likely to come into contact with a human and they have no hesitation in protecting themselves with their fangs and venom. Interestingly, it seems that defensive bites by *Phoneutria* wandering spiders often contain relatively small amounts of venom, which suggests these species may be able to control the amount of venom they use depending on the purpose of the bite.

Spiders of the genus *Loxosceles,* commonly known as recluse spiders, differ from the preceding genera in that they produce a venom that is largely necrotizing. Typically, these nondescript, drab spiders are very retiring, preferring to shelter in their messy webs in out-of-reach places rather than strutting about and showing off their venom. A recluse spider will only bite if handled or if it becomes trapped against the skin in clothing or bed linen. The component of recluse spider venom primarily responsible for killing cells and causing necrosis is an enzyme known as sphingomyelinase D. Initially, the bite is painless or causes a mild stinging sensation. However, this is the calm before the storm and after 2–8 hours painful blistering and swelling develop at the site of the bite. After about three days an ulcer has developed with a central portion composed of dead and dying tissue. The effect of the recluse spider venom on the tissues is to cause systemic symptoms (systemic loxoscelism), including joint pain, chills, fever, rash, nausea, and vomiting, followed by blood abnormalities, febrile seizures, coma, and acute renal failure in rare cases. If the bite wound is taken care of correctly, it heals over a 1–2 month period, although major scarring can occur in 10–15 percent of cases. From a 1997 U.S. sample of 111 patients with expert-confirmed brown recluse spider (*L. reclusa*) bites, no fatalities were reported and systemic loxoscelism symptoms were reported in just 3 percent of patients. In South America, systemic loxoscelism is more common, occurring in 13.1 percent of 267 patients bitten by *L. gaucho, L. laeta,* or *L. intermedia.* Of the patients bitten by the latter two species, 1.5 percent died.

An innate fear of spiders may have been advantageous to the survival of our ancestors, which is a possible explanation for why there are many people today who have an irrational fear of these arthropods—it may be hard-wired in our brains. However, as we have seen, only a few spiders are a cause for concern and even the very dangerous species have only caused a very small number of deaths in the last few decades. For those people who live in remote tropical areas and even those in less isolated areas who live in extreme poverty and cannot afford to pay for medical treatment, spider bites remain a danger, occasionally a life-threatening one. The true burden of spider bites is impossible to accurately assess because lesions caused by other organisms are commonly attributed to spiders, and even if a spider is involved, there is rarely a specimen with which to confirm an identification. Whatever the true incidence of morbidity and mortality attributable to spiders, we must remember that bites from these arachnids pose much less of a risk to humans than do the stings of insects such as bees, wasps, and hornets. These insects are considered to be relatively innocuous albeit commonly encountered animals, but in the United States alone they account for vastly more deaths than any other venomous animals. Between 1991 and 2001, 533 deaths were attributed to bees, wasps, and hornets, while only 66 deaths were attributed to spiders during the same period. It has been estimated that spider bites account for around 200 deaths each year around the world. However, spider bites cause morbidity other than death, ranging from pain lasting no more than 24 hours to significant, long-term, and even permanent injury. Morbidity resulting from spider bites is probably common in the rural tropics and is certainly underreported. In view of this, morbidity from spider bites may affect tens if not hundreds of thousands of people around the world each year, which translates into a considerable economic burden, compounded by poverty and other diseases.

Even in parts of the tropics and subtropics where spider bites are more common, it is relatively easy to reduce the risk of being bitten with some straightforward measures. Spiders seek out refuges in which to hide during the day and during the breeding season when females construct silken egg cases. Any material, in or around the home, that offers nooks and crannies for a spider to hide in should be cleaned up or moved. These include boxes, shoes, rarely worn clothes, and log piles in yards. In areas that may be inhabited by potentially dangerous spiders, care should be taken not to probe around blindly with the fingers, creating a situation in which a spider may bite defensively.

Finally, it's important to remember that spiders are integral parts of terrestrial ecosystems and that to limit the minimal damage they may cause us as a species we must respect their requirements as fellow beings.

FURTHER READING

Diaz, J. H. The global epidemiology, syndromic classification, management, and prevention of spider bites. *Am J Trop Med Hyg* 71(2)(2004): 239–50.

Foelix, R. F. *Biology of Spiders.* Oxford University Press, Oxford, United Kingdom, 1996.

Langley, R. L. Animal-related fatalities in the United States: An update. *Wilderness Environ Med* 16(2)(2005): 67–74.

Meier, J., and J. White. *Handbook of Clinical Toxicology of Animal Venoms and Poisons.* CRC Press, Boca Raton, LA, 1995.

Ticks

Ticks, like fleas and lice, are another group of superbly adapted ectoparasites that suck blood from a huge range of vertebrate hosts. These arachnids, closely related to the mites, are represented by around 870 species worldwide, but their relatively small size means there are probably many more species yet to be formally identified. It is thought these pesky invertebrates have their origins somewhere in the cretaceous period (65–146 million years ago) and that they underwent a radiation in diversity 5–65 million years ago. However, it is entirely possible these arachnids are considerably more ancient than these estimates suggests as the arthropod branch of life to which they belong was among the first of the all animal groups to leave the oceans to seek a life on land at least 400 million year ago.

The ticks are classified into three families: ixodidae (hard ticks—683 species), argasidae (soft ticks—183 species), and nuttalliellidae (1 species). As their names suggest, the first two families of ticks can be differentiated by the toughness of their bodies. The hard ticks have quite an armored appearance, while the soft ticks have a wrinkled abdomen (when unfed) that swells to accommodate the blood meal.

Ticks have a suite of adaptations to an ectoparasitic way of life. The limbs afford these arachnids an excellent purchase on their hosts and once on board they bring their mouthparts to bear to penetrate the skin and drink the blood of the host. The blood meals they consume are, relatively, probably the largest ingested by any bloodsucking animal. Tick larvae in the family ixodidae are able to consume 11–17 times their own pre-fed body weight in blood in one go, whereas adult females in the same family can pack away a truly remarkable 60–120 times their pre-fed weight in vertebrate blood. Astounding as their appetites may appear, the actual volume of blood ingested during each feeding event is even larger than

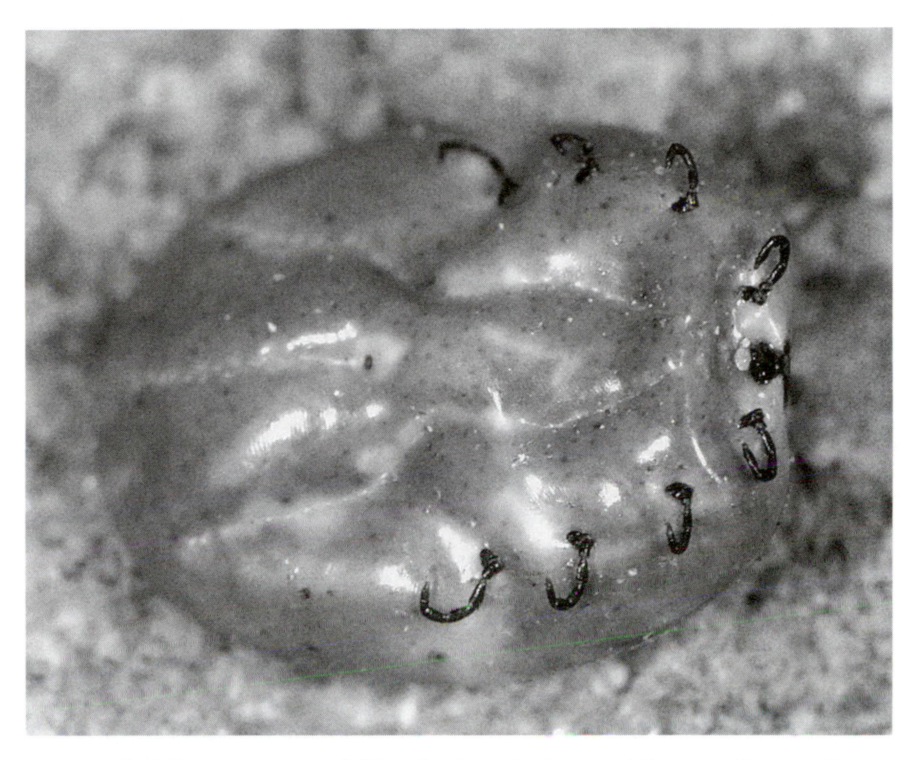

A soft tick fully engorged with blood. Note the legs and the mouthparts. The body of these animals can swell enormously to accommodate their blood meal. (Courtesy Ross Piper)

the figures given above because ticks are able to concentrate the blood meal by secreting unwanted fluids in the meal back into the host. Dealing with such a massive amount of food requires extensive morphological and physiological modifications. These adaptations allow the gut, body wall, and cuticle to expand sufficiently in a short space of time to accommodate the relatively prodigious volumes as well as quickly ridding the body of the huge surfeit of water and salts.

The life cycle of hard ticks begins with the production of large numbers of eggs. Fecundity varies according to species, but typically, female hard ticks produce a few hundred to more than 10,000 eggs; however, there are observations of the females of some species (*Amblyomma variegatum*) producing 34,000 eggs. This is a phenomenal number of eggs for a terrestrial invertebrate and it says an awful lot about the chances of the young ticks finding a host to feed from before they use up their energy reserves and starve. As soon as the larvae hatch they can go about getting to grips with a host, but because they have neither a good turn of speed or wings they rely on a suitable host wandering sufficiently close for them to latch on. Once on board they rasp their way through the host's skin and begin feeding. Following their first blood meal the larvae of the majority of hard tick species drop from their host to digest their food and metamorphose into a nymph. The nymph then goes about finding another host, feeds on it, drops off, digests its food, and molts to give rise to the adult hard tick. Yet again, in the third and final stage of life, the ticks are faced with the challenge of finding a host. If they're successful, they feed and mate, and then the fully engorged female drops off to digest the massive amount of food she has just ingested as well as to lay her eggs. With her eggs laid, the female has done her job and she dies. From larva to adult, hard ticks spend the vast majority of their time away from their hosts; and each stage in the life cycle of larva, nymph, and adult may utilize different host species. In northern latitudes, the complete life cycle of the hard ticks can be as long as three years as each stage takes about a year to complete.

The life cycle of the soft ticks is slightly different in that there are two to seven nymphal instars. Again, each stage in the life cycle requires a blood meal, often from different host species. Instead of imbibing a single huge blood meal, adult female soft ticks take a number of smaller meals and after each one they drop from their host, digest their food, and produce a small batch of eggs, the size of which varies according to the species, but is normally around 500. There can be as many as six of these feeding/egg-laying events with each one separated by an interval of several months, often extended by the ability of the ticks to go without food for months at a time. This is especially true for those soft tick species dependent on migratory vertebrates. These species are able to survive without food for several months by entering a state of suspended animation (diapause) where their metabolic rate falls to almost immeasurable levels in order to conserve energy.

Sucking blood from a vertebrate is not without its difficulties. Apart from the risk of getting squashed or brushed off there is the battery of defenses protecting the animal's body once the outer wall, the skin, has been breached. Any bloodsucking arthropod needs the blood to flow freely from its host, but as soon as a blood vessel is ruptured a cascade of events begins to close the wound. The saliva produced by ticks is perhaps the most sophisticated concoction produced by any blood-feeding parasite. This saliva contains a number of factors to prevent blood clotting and the formation of new blood vessels, both of which are crucial in the wound-healing process. The cocktail also contains compounds to numb pain, itching, and inflammation, potential giveaways to a host that something is amiss when it can't see or reach an ectoparasite on its body.

Of the animals that inadvertently inflict misery on humans and domesticated animals, the ticks are second only to the mosquitoes and in some areas of the world these arachnids are the most important. Their blood feeding activities have four important implications for human and animal health: blood loss, dermatosis, paralysis, and disease. Blood is not a problem when there are only one or two ticks feeding from a host, but when the infestation is large—several hundred ticks—the amount of blood consumed outstrips the host's ability to replace it and anemia can develop. It is not uncommon for large host animals, such as cattle with heavy tick infestations, to lose 80–90 kilograms of blood in a single season thanks to the relatively massive appetites of these parasites. The bites of ticks can cause a number of problems in the epidermal tissues of their hosts, including inflammation, swelling, itching, and ulcerations, which can be particularly severe in favored feeding places, such as the ear canals of animals like dogs and cats. These problems can arise as a direct result of the mechanical damage caused by the tick's mouthparts or the substances in its saliva. Dermatosis can be further compounded by secondary infection of the wound by bacteria, which in some cases can lead to serious complications. The chemical cocktail that makes up the tick's saliva also contains certain compounds that can act like toxins and when these are injected near the base of the host's skull they can cause paralysis in humans, cattle, dogs, and other mammals. The paralysis isn't permanent and it can be quickly reversed by removing the tick. Blood loss, dermatosis, and paralysis are all relatively minor concerns compared to the varied and often serious tick-borne diseases caused by viruses, bacteria, and protozoa that affect humans and domesticated animals all over the world. Some of the most important of these diseases are shown in the sidebar.

To go into much depth on the diseases above and the others transmitted by ticks would require a book in itself, so we'll focus on the costs of some of these diseases and the impact they have around the world. Until fairly recently, tick-borne diseases of humans were on the decline or were not considered to be of much concern, but there has been a worrying increase in the reports of these diseases over the last few years. One of the diseases of the most concern is Lyme disease, which is not a problem if treated early, but can cause extremely debilitating symptoms if left untreated. In the United States in 1982, 497 cases of Lyme disease were reported, but between 2003 and 2005, 64,382 cases were reported, an average of more

Some of the Important Tick-borne Pathogens, the Diseases of Humans and Domesticated Animals They Cause, and the Tick Species Responsible For Their Transmission

	Humans*	
Pathogen	**Disease**	**Vector tick(s)**
Viruses		
Flavivirus	Tick-borne encephalitis	*Ixodes ricinus, I. persulcatus*
Coltivirus	Colorado tick fever	*Dermacentor andersoni*
Flavivirus	Powassan encephalitis	*Ixodes, Dermacentor,* and *Haemaphysalis* spp.
Bacteria		
Borrelia burgdorferi and other *Borrelia* spp.	Lyme disease	*Ixodes ricinus* and other *Ixodes* spp.
Ehrlichia ewingii, E. chaffeensis	Human ehrlichiosis	*Amblyomma americanum*
Anaplasma phagocytophilum	Human anaplasmosis	*Ixodes ricinus* and other *Ixodes* spp.
Rickettsia rickettsii	Rocky Mountain spotted fever	*Dermacentor variabilis* and other *Dermacentor* spp.
Borrelia spp.	Tick-borne relapsing fever	*Ornithodoros* spp.
Francisella tularensis	Tularemia	Many species
Coxiella burneti	Q fever	Many species
Protozoa		
Babesia microti, B. divergens, B. duncani, B. venatorum	Babesiosis	*Ixodes scapularis* and *I. ricinus*

	Domesticated animals	
Pathogen	**Disease**	**Vector tick(s)**
Viruses		
Iridovirus	African swine fever	*Ornithodorus porcinus*
Flavivirus	Louping ill, a.k.a. ovine encephalomyelitis	*Ixodes ricinus*
Bacteria		
Anaplasma marginale, A. central, A. ovis	Anaplasmosis	*Dermacentor* spp., *Rhipicephalus* spp., *Hyalomma* spp.
Borrelia burgdorferi	Borrelioses	*Ixodes* spp.
Ehrlichia canis, E. ewingii	Ehrlichiosis	*Rhipicephalus sanguineus, Ixodes ricinus, Amblyomma americanum,* and others
Ehrlichiosis ruminantium	Cowdriosis, a.k.a. heartwater disease	*Amblyomma* spp.
Protozoa		
Theileria parva	East coast fever	*Rhipicephalus appendiculatus*
Theileria annulata	Tropical theileriosis	*Hyalomma* spp.

*Many of the tick-borne pathogens causing disease in humans also cause disease in domesticated and wild animals and vice versa.

than 20,000 a year, which represents an increase of around 4,000 percent. Exactly why this disease is on the rise is a bone of contention, but it is probably due to a number of factors, including the encroachment of development into wild habitats, more people taking part in outdoor leisure activities, greater awareness of the dangers of Lyme disease, and the effect of climate change on the range and behavior of the vector ticks. Globally, it has been estimated that there are at least 100,000 cases of human illness every year related to tick-borne diseases, but the real figure is undoubtedly far higher. Many people at risk from these diseases inhabit remote rural areas where adequate medical facilities are lacking, problems that are compounded by the difficulty in diagnosing tick-borne diseases.

The current public health impact of ticks and the diseases they transmit is minor compared to the havoc they wreak in agriculture. The full extent of the tick and tick-borne disease problem in farming can only really be appreciated in the developing world where these invertebrates are a major impediment to agricultural advancement and food independence, which is a complex ethical and environmental topic in its own right (see introduction—pests and the environment section). It has been estimated that the total global losses attributable to ticks and the diseases they transmit is somewhere in the region of $13.9–18.7 billion, but as with any estimate of the cost of such an enormous problem it is probably well short of the true amount. In addition, it is thought that at any one time at least 800 million cattle around the world are continuously exposed to ticks and tick-borne diseases. In Tanzania alone, the estimated annual loss to the cattle industry accounted for by tick-borne disease, specifically anaplasmosis, babesiosis, cowdriosis, and theileriosis, is $364 million. Theileriosis accounted for 68 percent of this loss, while anaplasmosis, babesiosis, and cowdriosis were responsible for 13 percent, 13 percent, and 6 percent, respectively. A loss of this magnitude would be a real problem anywhere in the world, but in the developing world this is a disaster. This cost includes the death of around 1.3 million cattle, infection, treatment, as well as milk and weight loss.

All major attempts at controlling ticks have been largely unsuccessful for a number of reasons. These include the ability of ticks to quickly evolve resistance to acaricides, the behavior of dropping from a host after a blood meal, and the long periods of time they spend away from their host digesting food, molting, and reproducing. With control being ineffective, the best way of limiting the impact of these arachnids and the diseases they transmit is preventing them from biting in the first place. In some situations, such as the backyards of houses and workplaces in developed nations, undergrowth and other vegetation that ticks use as perches to clamber onto a passing host can be simply cut back and maintained. Long clothing and repellents can prevent the ticks from getting to the skin and biting. Light-colored clothes allow the ticks to be easily seen and plucked off, and tucking trousers into socks prevents ticks from getting onto the skin. Simple measures such as these can be very effective in stopping these parasites from biting and then transmitting diseases. In developing countries, the resources are not often available for the adoption of simple preventative measures and because agriculture in these countries is often far less intensive than in the developed world, suitable habitat for ticks is very abundant. The typical strategy in these areas is the widespread use of acaricides. Tanzania and Uganda each spend around $26 million every year importing these chemicals. Not only are these chemicals hazardous to the environment, but as has already been mentioned, resistance to them evolves very rapidly, rendering them close to useless.

The impact of ticks on human and animal health is undisputed, but it's useful to take a look at animals like the ticks and reflect on how successful they are. Evolution has honed these little arachnids into perfect parasites and they're only really a problem because of the ever-increasing pressure that humans are placing on the environment in the requirements for more land to build and farm on. Ticks are simply doing what they have always done—making a difficult living on larger animals. It would be better to find ways to live alongside them instead of making futile and damaging efforts at controlling them.

FURTHER READING

Barker, S. C., and A. Murrell. Phylogeny, evolution and historical zoogeography of ticks: A review of recent progress. *Experimental and Applied Acarology* 28(2002): 55–68.

Fuente, J. The fossil record and the origin of ticks (Acari: Parasitiformes: Ixodida). *Exp. Appl. Acarol* 29(2003): 331–44.

Goodman, J. L., D. T. Dennis, and D. E. Sonenshine. *Tick-borne Diseases of Humans.* ASM Press, Washington, DC, 2005.

Gubler, D. Resurgent vector-borne diseases as a global health problem. *Emerging Infectious Diseases* 4(3)(1998): 442–50.

Kivaria, F. M. Estimated direct economic costs associated with tick-borne diseases on cattle in Tanzania. *Tropical Animal Health and Productivity* 38(2006): 291–99.

Nicholson, W. L., D. E. Sonenshine, R. S. Lane, and G. Uilenberg Lloyd. Ticks (Ixodida). In *Medical and Veterinary Entomology* (G. R. Mullen and L. A. Durden, eds.), pp. 493–542. Academic Press, San Diego, CA, 2009.

Varroa Mite

The mite *Varroa destructor* is a honeybee parasite with a global distribution that is responsible for economic losses amounting to hundreds of billions of dollars each year. No more than 1–2 millimeters in size, this tiny, rather crab-like arachnid was first identified from Southeast Asia in 1904. Since then it has been inadvertently introduced to countries all around the world with the first U.S. infestations being recorded in 1987 (Wisconsin and Florida). The mite is a specialized ectoparasite of a number of bee species, which develops on the brood of these social animals, necessitating a life cycle closely synchronized to the host. The female mite lays her eggs on the developing honeybee larvae and the young mites (typically several females and one male) feed on the bee's hemolymph (the insect equivalent of blood). The mites develop into adults and the male mates with all his sisters, fertilizing their eggs. Soon after, his job complete, the male dies. Eventually, the bee larva pupates and the female mites cling on to their changing host because this is the only way they will escape the brood cell and get access to other immature bees to find food for their own offspring.

The predilection for hemolymph by *Varroa* mites is one way in which these creatures can be injurious to bees, but increasingly, the negative effects of an infestation result from several viruses transmitted by these arachnids, including deformed wing virus, acute bee paralysis virus, and slow paralysis virus. The feeding activities and the diseases transmitted by the mites have dire consequences for individual bees and the hive as a whole. The transmission of viruses and the loss of a relatively large amount of hemolymph from an immature bee are compounded when the larva pupates because this life stage is very sensitive to damage and disease. Should the developing bee be weakened or diseased the resultant adult is often a very sorry specimen with deformed wings and abdomen. A small *Varroa* infestation may not be a problem for the hive as a whole, but in the case of a heavy infestation lots of potential workers are damaged and the colony may cease to function, eventually leading to its collapse. Recently, a worrying phenomenon termed *colony collapse disorder* has been observed around the world, whereby bee colonies mysteriously die off, leaving huge numbers of hives empty and commercial apiculturists massively out of pocket. The exact cause of this phenomenon is unknown, but it is very likely that a number of factors are to blame, one of which is the tiny *Varroa* mite.

A magnified view of a *Varroa* bee mite. *Varroa* mites are external honeybee parasites that attack both the adults and the brood. (Dennis Kunkel Microscopy, Inc./Visuals Unlimited, Inc.)

The economic toll of the *Varroa* mite is immense, which may be hard to believe because all that most people associate with bees is honey. Honey is just one of the things these amazing insects produce. Other products of the honeybee's industrious activities include things like wax, propolis, and royal jelly, all of which are used in a variety of human industries. However, the damage to the bee product industry pales into insignificance compared with the huge impact declining honeybee populations has on the pollination of crops and wild plants. A huge number of crops rely on the honeybee for pollination (see list below) and without honeybees to do this job we would be without such staples as almonds, onions, and apples, to name but a few. It is very difficult to quantify exactly how much the pollination services provided by honeybees are worth. Estimates in the region of $217 billion have been suggested, but it is likely the real figure is many times greater.

Crops in Which Honeybee Pollinators are Important or Crucial				
Alfalfa	Chinese	Dewberry	Muskmelons	Plums &
Allspice	gooseberry	Drug plants	Cantaloupe	prunes
Almonds	or kiwi	Eggplants	Casaba Crenshaw	Pumpkin &
Alsike clover	Cicer milkvetch	Garlic	Honeyball	squash
Apples	Cinnamon	Gooseberries	Honeydew	Quinine
Avocado	Citron	Herbs (spices)	Persian melon	Radish
Berseem	Citrus	Huckleberry	Mustard	Rape
Blackberries	Pummelo	Jujube	Niger	Raspberries
Blueberries	Tangelo	Kenaf	Nutmeg	Red clover
Buckwheat	Tangerine	Kohlrabi	Parsley	Rutabagas
Cacao	Clovers, minor	Kola nut	Parsnip	Sainfoin
Carambolo	Cranberries	Lavender	Passion fruit	Sapote
Cardamom	Crimson clover	Litchi	Peaches &	Sunflower
Cashew	Crownvetch	Longan	nectarines	Sweetclovers
Celeriac	Cucumbers	Lotus	Pears	Sweetvetch
Chayote	Currants	Macadamia	Persimmon	Tea
Cherries	Cutflower seeds	Mango	Pimenta	Trefoils

The stakes are high and one factor in the decline of the honeybee, the burgeoning *Varroa* mite population, is proving very difficult to control. Limiting the spread of this parasite has met with moderate success in some areas, but even in places as geographically isolated as New Zealand, the mite has managed to invade and wreak havoc on the bee and pollination industry. Acaricides are available, but overzealous use of these chemicals has led to the rapid evolution of resistance in the target mites. Various biological control agents are currently being investigated and one of the most promising of these is a parasitic fungus that infects and kills these mites. It has also been observed that some types of honeybee are able to rid themselves of these mites by grooming. Crossing honeybee types to select for these behavioral traits combined with biocontrol and the judicious use of pesticides will offer the greatest hope for controlling the *Varroa* mite.

FURTHER READING

Avitabile, A., D. Sammataro, and R. Morse. *The Beekeeper's Handbook.* Cornell University Press, Ithaca, NY, 2006.

Fernandez, N., Y. Coineau, and P. Theron. *Varroa: Serial Bee Killer Mite.* Atlantica-Séguier, Paris, 2007.

Mobus, B., and L. J. Connor. *Varroa Handbook: Biology and Control.* Northern Bee Books, Hebden Bridge, United Kingdom, 1988.

Crustaceans

Fish Lice

Fish lice look like something that crept from the fertile imagination of a science fiction writer. Flattened, disc-shaped, and equipped with suckers, claws, and a cruel-looking stylet, these bizarre animals are actually small, specialist crustacean parasites of fish. Many species are known from all around the world, but the most important species are those in genus *Argulus* as they cause considerable economic damage to fish-farming ventures the world over.

Argulus fish lice begin their life as eggs deposited by their mother on submerged surfaces, such as rocks. As soon as they hatch the young crustaceans are active swimmers—they have to be, because if they don't locate a host in as little as two days their energy reserves will be exhausted and they'll die. Vision appears to be the most important sense in host location and the compound eyes of these parasites are well developed, enabling them to find fish in murky water. Host location is not much of a problem in fish farms because the fish are so densely packed, but fish lice in the wild probably have a very difficult time locating their hosts. Once the *Argulus* has found a suitable host it attaches to the unfortunate victim's body with its modified mouthparts—its suckers, which are considered to be among the most elaborate in the animal kingdom. Once securely attached, *Argulus* begins to rasp at the flesh with its mandibles, sucking edible particles into its stomach. *Argulus* are found on the body of the fish and also beneath the gill covers. If they are disturbed or feel like looking for a new place to feed they are able to wiggle rapidly over the fish by using their big suckers alternately.

Depending on temperature and with access to a consistent supply of food, the *Argulus* can reach maturity in around four weeks. At this point breeding can take place—often on the host. Reproduction in these crustaceans is very interesting as the male has no penis, neither does he produce any form of spermatophore. Close observations have revealed that the male and female lock themselves together using various spiny structures on their bodies and ducts containing the male's sperm are pierced

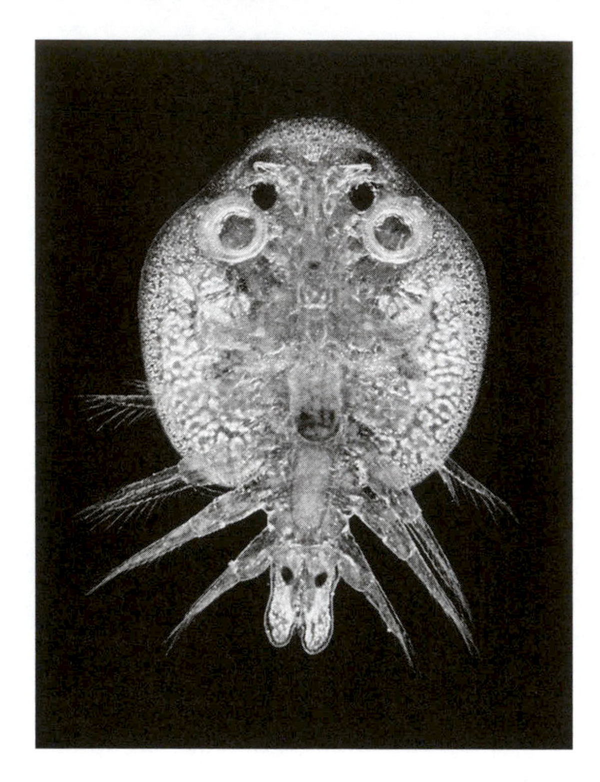

The fish louse (*Argulus*) is equipped with two suckers that enable this flat freshwater crustacean to attach itself to its fish host. (Wim van Egmond/Visuals Unlimited, Inc.)

by specialized spines of the female's reproductive apparatus. The sperm in these ducts is at high pressure and it flows out into the female's genital tract to fertilize her eggs.

Interesting life history aside, these crustaceans are a serious pest of freshwater fish farms. The most obvious problem associated with *Argulus* is the stress and blood loss caused by their feeding activity. If one or two *Argulus* are attached to a host the effects are negligible, but heavy infestations can make the fish very lethargic and stop it from feeding. A more serious consequence of *Argulus* infestations are the microorganisms that gain entry to the fish via the wounds caused by the parasite's feeding activity. These secondary infections can often be fatal. In the densely populated environment of the fish farm they can be devastating. In addition, fish lice are also a vector for diseases, including those caused by viruses, such as spring viraemia of carp and conditions caused by parasitic nematodes.

Important commercial fish species parasitized by *Argulus* and other fish lice include salmon, trout, and carp. These species constitute a considerable proportion of the global aquaculture industry, estimated in 2005 to be worth $78.4 billion. China alone has by far the largest aquaculture industry in the world and in 2005 inland freshwater ponds produced more than 14 million tonnes of carp destined for human consumption. The propensity of *Argulus* to feed on freshwater fish levies a heavy economic toll on the aquaculture industry, costs that can be broken down into restocking, yield reduction, and monies associated with controlling these parasites and treating the consequences of their activities. Quantifying this economic cost is very difficult, but globally it must certainly amount to hundreds of millions of dollars annually.

Controlling *Argulus* fish lice once they have become established in a fish-farm situation is not easy and measures are limited to preventing heavy infestations rather than eradication. Pesticides are routinely applied by dissolving them in the infested water; however, the side effects of these chemicals on ecosystems are impossible to ignore and research is ongoing to identify more environmentally friendly ways of controlling these pests. One successful, albeit time-consuming way of controlling *Argulus* is good animal husbandry and stock management, including the inspection of individual fish for these relatively large parasites, which can then be removed with forceps before being unceremoniously disposed of.

FURTHER READING

Avenant-Oldewage, A., and J. H. Swanepoel. The male reproductive system and mechanism of sperm transfer in *Argulus japonicus* (Crustacea: Branchiura). *Journal of Morphology* 215(1993): 51–63.

Kearn, G. C. *Leeches, Lice and Lampreys: A Natural History of Skin and Gill Parasites of Fishes.* Springer, Dordrecht, Germany, 2004.

Gill Maggots

It's hard to imagine animals more ghastly, more distasteful to the eye than some creatures that parasitize fish. I for one never cease to be amazed at how the blind power of evolution can shape an organism to be infinitely more disturbing than the most infamous creations of horror and science fiction.

We've already seen the fish louse, but another, even more repugnant-looking, albeit interesting, parasite of fish responsible for considerable economic losses are the gill maggots. Contrary to their common name, these creatures are not maggots, but they do live on gills. They are actually free-loading copepods in the genus *Ergasilus* and unlike many of the

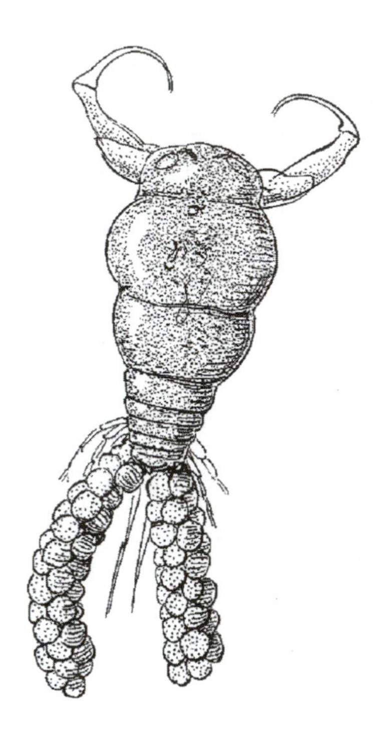

Ergasilids, commonly known as gill maggots, are important parasites of fish. They cling on to the host's gills using their highly modified antennae. The female in this image has a pair of egg masses attached to her abdomen. (CDC)

parasitic copepods they retain many of the primitive features characterizing this group of minute crustaceans. For instance, they retain two pairs of antennae, completely absent in the more derived parasitic copepods, but as a concession to their parasitic way of life the long second pair of antennae are greatly modified to form a pair of grasping appendages used to great effect by gill maggots to anchor themselves to the slippery gills of their hosts. They also retain some degree of swimming ability and are quite able to propel themselves through the water should they become detached from their host.

After hatching from their eggs, the gill maggots go through a number of larval stages, many of which are spent in the aquatic soup known as plankton, where they feed on the other microscopic representatives of the planktonic community. Upon reaching adulthood, the gill maggots mate and the female swims off to search for a host, while the adult male remains a free-living creature in the plankton. In the vast majority of cases, the adult female gill maggots may never chance upon a host and they will perish. However, for a fortunate few, a suitable host may swim past and they grasp the opportunity literally, swimming under the gill cover and latching onto one of the gill filaments with the cruel-looking claw at the end of the highly modified second antennae. In some species of gill maggot, the clawed tips of the antennae actually fuse, completely encircling the gill filament and providing an almost unassailable grip for the entire life of the parasite. Firmly fixed to the fish's gas-exchange apparatus, the female gill maggot goes about feeding. Her first pair of legs are adorned with blade-like spines and it is these she uses to rasp mucus and tissue from the poor host's gills into her waiting maw, a process aided by the secretion of digestive enzymes. On this protein- and lipid-rich diet the female gill maggot's eggs mature, borne in two elongated sacs attached to her hind end. Depending on the species in question as well as many other factors, each adult female gill maggot can produce 20–100 eggs and it is from these the free-living larvae hatch to complete the life cycle.

The parasites are a problem because a host fish is normally inhabited by many gill maggots, all of which are latched onto the fish's gills and rasping at the delicate tissue. The physical attachment of the gill maggots and their feeding activities damage the gill tissue, interfere with gas exchange, and cause wounds that permit the entry of pathogens. All of these ills may eventually cause the death of the fish. Where freshwater or certain marine fish are farmed the gill maggots can be a huge problem, because the crowded conditions make it very easy for the parasites to find hosts. Tilapia, carp, and mullet fisheries are known to be vulnerable to the depredations of the gill maggots. In these situations it is not unusual for an individual fish to be infested with thousands of these crustaceans. With such heavy infestations it is not surprising that fish farming losses can be considerable. The large group of commercially farmed cichlid fish, collectively known as tilapia, can be severely affected by gill maggots. More than two million tonnes of one cichlid species—the Nile tilapia (*Oreochromis niloticus*)—were produced globally in 2007 and in many developing countries these fish represent the only source of high-quality animal protein available to a large proportion of the population. Therefore, any organism that causes losses in the tilapia farming industry can have important consequences for the health and welfare of people throughout entire regions. In Israel, where tilapia are cultured on a large scale, heavy infestations of gill maggots have been estimated to cause harvest losses of around 50 percent. If losses of this magnitude are experienced in Israel, an affluent country with the means at its disposal to control such pests with commercially available poisons, then the losses in poorer areas where these fish are farmed, such as sub-Saharan Africa, must be huge.

In Israel, gill maggot infestations have been treated with bromhexine hydrochloride, a compound used in medicine to treat disorders of the lungs by dissolving excess mucus. Dispensed into the water this compound is known to be very effective in controlling gill maggot numbers, but such a treatment is beyond the means of many people who farm fish in developing regions. Organophosphate pesticides have also been used to treat gill maggot infestations with varying degrees of success; however, their effect on nontarget organisms and accumulation in the food chain means they are detrimental to both the environment and human health.

FURTHER READING

Roberts, L. S., and J. Janovy, Jr. *Foundations of Parasitology.* McGraw-Hill Higher Education, NY, 2008.

Woo, P.T.K. *Fish Diseases and Disorders: Vol. 1. Protozoan and Metazoan Infections.* CAB International, Wallingford, United Kingdom, 1995.

Insects

Aphids

Aphids, the small squidgy animals familiar to anyone with houseplants or a garden as *greenfly* (something of a misnomer) are perhaps the most important insect pests of agriculture. The 4,400 or so species are found around the world, but are most abundant in the temperate regions of the northern hemisphere. They feed on all types of plants, including trees, shrubs, herbs, and grasses. Some species will feed on a large number of host plants, while others are more selective and are associated with one species of plant. They are believed to have appeared over 280 million years ago when there were far fewer plant species than there are today. Around 100 million years ago, there was an explosion in the variety of flowering plants and the aphids diversified to exploit this new abundance of food.

Although abhorred by farmers and gardeners the world over, the aphids are amazing little animals, perfectly adapted to take advantage of the way in which humans have manipulated the environment to grow food. In this respect they are extremely successful animals. It's worth understanding a little bit about the biology of these sapsuckers, and in doing so you may look at these pests in a slightly different way the next time you see some of them.

Aphids are all sapsuckers. Their piercing mouthparts are like a feeding straw, the tip of which secretes a fluid that hardens to form a tube. Their aim is to pierce the phloem vessels transporting sap around the plant, but even when they manage this the plant rallies its defenses to plug the holes. To fool the plant's defenses, aphids produce a number of proteins that disguise its activities, allowing it to breach the phloem and suck the sap. Sap may be sugary, but other important nutrients are only present in minute quantities, so aphids imbibe large quantities of it and enlist the help of symbiotic bacteria and yeasts to digest the sap as efficiently as possible. The excess water and sugar is rapidly processed in the aphid equivalent of kidneys to emerge at the hind end of the animal as what is commonly known as honeydew.

Aphids are commonly tended and protected by ants because of the sweet honeydew they secrete as a by-product of their sap-sucking. (Courtesy of Ross Piper)

Lots of insects suck sap in one way or another, so in this respect the aphids don't really stand out; however, what does single them out is their reproductive ability. Any gardener will know that in a very short amount of time, seemingly overnight, a plant free from aphids can be swarming with them. For much of the year, many species of aphid reproduce without mating, a process that begins with a female who hatched from an egg laid in a suitably secluded spot, such as the deep fissures in tree bark, the previous year. This founding female had a mother and a father, but the odd makeup of the aphid's chromosomes means that a mating between a male and female can only produce daughters. These daughters survive the winter and within them they carry the seed of the new season's population. The founding female is already carrying a daughter and within this embryo another embryo develops; three generations in the body of one tiny animal all produced via the process of parthenogenesis—reproduction without sex. These daughters are born as miniature replicas of their mother and they too give birth to further replicas until there are huge numbers of aphid, all originating from the original female who survived the winter as an egg. The reproductive capacity of aphids is astounding. A single cabbage aphid (*Brevicoryne brassicae*) can produce more than 40 generations of females in a single season and if all of them survived we would be knee-deep in these tiny creatures by the fall.

During the autumn the aphid colony will start producing males and females whose function it is to mate and produce the founding female for the following year. In certain species of aphid, some of the clones, although genetically identically to the original female, will look slightly different and perform certain tasks, such as guarding the colony. These castes are commonly soldiers with enlarged front legs and a spiky head used to jab and prod animals that threaten the colony. During the feeding season the aphid colony may become too big, resulting in overcrowding that may kill the host plant. In these situations the aphids start giving birth to winged individuals. These alates, as they are known, will leave the colony to search for new food plants.

Aphids are a problem for a number of reasons. Firstly, they drink the plant's sap, which is needed to fuel the growth of leaves, buds, and flowers. Also, and probably most important of all, aphids are vectors of plant diseases. They breach the outer defenses of plants with their mouthparts, transmitting numerous pathogens, many of which are viruses that can devastate whole crops. Of all the known plant viruses, slightly more than half are transmitted by aphids. Lastly, their reproductive potential and mutualistic relationships with other animals, notably ants, allows them to quickly build and maintain large populations. Some of the more important pest aphids can inflict heavy losses on crops (see sidebar).

Aphids are so widespread and feed on so many crops that it's impossible to put an exact figure on the economic damage they cause. Between 1986 and 2001, the Russian wheat aphid is estimated to have cost U.S. farmers around $1 billion in yield losses and control. Every year in certain parts of the United States the green bug (*Schizaphis graminum*) accounts

Some Important Pest Aphids and Their Impact on Crop Yields

Aphid species	Crop affected	Yield reduction
Pea aphid (*Acyrthosiphon pisum*)	Pea	~16%
Cowpea aphid (*Aphis craccivora*)	Ground nut	>50%
Peach potato aphid (*Myzus persicae*)	Potato	5–30%
Black bean aphid (*Aphis fabae*)	Field beans	>50%
Russian wheat aphid (*Diuraphis noxia*)	Wheat	>50%
Corn aphid (*Rhopalosiphum maidis*)	Corn	~36%

for losses of $12 million, rising to $100 million in years when there are severe outbreaks.

Farmers and gardeners have a range of ways to control aphids. By far the most common is the use of insecticides. These chemicals do kill the aphids, but their effects on beneficial organisms and the wider environment can be devastating. The other major problem with insecticides is resistance, which renders these chemicals useless in the long term. Apart from chemical control, aphids have plenty of natural enemies, including many predators, parasites, and pathogens. Some of these can be harnessed as biological control agents either by breeding them en masse and releasing them in places where they can't simply fly away, such as glasshouses, or by adopting agricultural practices that enhance the wild populations of these enemies, such as minimal tillage and preservation of noncultivated field boundaries.

FURTHER READING

Van Emden, H. F., and R. Harrington. *Aphids as Crop Pests.* CABI, Oxford, United Kingdom, 2007.

Asian Long-horned Beetle

This large, handsome beetle is a native of China and Korea, but in the 1980s it somehow found its way to North America and since then it has become an important invasive pest of a number of tree species, often causing the death of the infected trees. More recently, it has also found its way into Europe, specifically Austria, France, and Germany.

In China, the preferred host trees of this beetle are poplar trees (*Populus* spp.), willows (*Salix* spp.), elms (*Ulmus* spp.), and maples (*Acer* spp.), but it is also known to attack representatives of several other tree genera. In recent decades its range in China has greatly increased thanks to the large-scale planting of poplars and willows that are used for timber and timber products because of their rapid rate of growth. Like all beetles in the family cerambycidae (commonly known as long-horned beetles because they have very long antennae), the larvae of the Asian long-horned are specialized wood feeders. Each adult female beetle is capable of producing between 30 and 120 eggs depending on which host tree she feeds. Each of the eggs is deposited individually in a small niche the female makes in

A native of China and Korea, the Asian long-horned beetle has been accidentally introduced into North America and Europe. (U.S. Fish and Wildlife Service)

the bark of the host tree with her mandibles. These oviposition niches are typically made on the eastern side of the trunk or on branches with a diameter of more than 5 centimeters. The eggs take about two weeks to hatch, at which time the first instar larvae chew their way through the bark using their powerful mandibles and begin feeding on the cambium—the vessel-laden tissues of the tree that convey fluids and nutrients. After some time feeding in the outer layers of the wood, the larvae tunnel deeper into the heartwood, feeding on the wood as they go. After many months of feeding on the woody tissue of their host they pupate in a small chamber. The adult that emerges spends around seven days in the chamber before it has to chew its way out of the only environment it has known. To facilitate its escape from its brood tree, the adult is equipped with powerful mandibular muscles and extremely tough mandibles. After some hours or days of chewing the adult beetle reaches the outer reaches of the host tree and escapes through a perfectly formed, 10-millimeter hole.

The adults are powerful flyers and they are capable of dispersing more than a kilometer from their brood tree during their brief adult existence to seek out mates and new hosts. The adults feed on the leaves, petioles, and twigs of their host plants and they appear to attack both healthy and diseased trees. Between bouts of feeding the females mate, sometimes with a number of males, and the life cycle of this insect is perpetuated. The latitude at which these beetle are found determines the number of generations there are per year. In Taiwan, at the southern limit of the range of this beetle, there is one generation per year; however, the farther north they are found, the longer the larvae take to develop. In the north of China, a single generation of Asian long-horned beetles takes two years to develop.

The impact of this species has been considerable, not only on natural forests, but also on commercial plantations, urban trees, and parks. The damage caused by the feeding larvae can be severe enough to weaken and kill the tree, especially when they are in the cambial layers of their host. The tunneling larvae damage the phloem and xylem vessels, resulting in heavy sap flow from wounds, which can serve as points of entry for other pests and pathogens. From the loss of fluids, the infested trees lose turgor pressure and their leaves become yellow and droop. Tree death is usually slow and can take as long as three to five years. In addition to killing trees, the damage the larvae cause as they eat their way through the wood also considerably reduces the commercial value of timber.

In its native China, the Asian long-horned beetle has been responsible for damaging 40 percent of the country's poplar plantations—approximately 2.4 million hectares. Urban trees in 240 cities throughout five provinces have been attacked by this beetle and during a three-year period alone in Ningxia Province more than 50 million infested trees had to be cut down. Cutting the trees down may seem extreme, but in heavy infestations, infested branches and even whole trees can fall without warning, causing damage to property and injuring people and livestock.

Exactly how this species found its way to North America is not known, but it is presumed that wood used to construct packing crates and other containers for international trade inadvertently harbored the immature stages of the beetle or quiescent adults. The beetles would have completed their development in these relatively thin pieces of wood to emerge stateside as adults with an abundance of hosts at their disposal. Since the original detection of this species in North America its spread has been rather slow and limited to the eastern side of the country. As of May 2001, 5,286 infested trees in New York and 1,547 in Chicago have been cut down to limit the damage caused by falling branches and to slow the spread of this pest. Although the impact of this beetle in North America has not been huge to date, the potential exists for it to cause massive damage in the forests of the United States. It has been estimated that the Asian long-horned beetle has the potential to cause the loss of 71 billion trees with a value of $2 trillion and to cause further loss amounting to $669 million through damage of otherwise healthy trees. These figures represent nothing more than an estimate of the damage this beetle could cause if it spreads throughout the United States.

To date, the impact of this species in Europe has been minor, but the ease with which this beetle can be inadvertently transported, its dispersal ability as an adult, and its freedom from natural enemies in areas beyond its natural range means that only a small population is needed to initiate a continental outbreak.

Controlling this beetle is a challenge, because the immature life stages leave few clues to their presence, so the full extent of an infestation can be hard to gauge. Targeting the larvae and pupae with conventional pest control agents is also of limited use because they are concealed beneath many centimeters of wood that protects them from insecticide sprays. Injecting insecticides into the tree is one way of controlling infestations, but over large areas this treatment would be expensive and logistically difficult. Other, more environmentally friendly forms of chemical control are also being investigated, including pheromone traps that emit volatile compounds mimicking those produced by the beetles to find their conspecifics for the purposes of mating.

Biological control agents are also being investigated as a means of controlling these beetles. The wood-feeding behavior of long-horned beetle larvae protects them from myriad predators, but there are many parasitic organisms that have evolved to prey on insect larvae developing deep inside trees, or on the eggs and very young larvae that have not had a chance to tunnel into their host plant. Currently, a number of fungi, nematodes, bacteria, microsporidia (tiny intracellular parasites related to fungi), and other insects have been identified that kill the immature stages of the beetle.

Of particular interest as a potential biological control agent is the cylindrical bark beetle (*Dastarcus longulus*), which parasitizes the larvae and pupae of the Asian long-horned beetle as well as other related beetles. In the native range of the Asian longhorn, *D. longulus* reportedly kills around 60 percent of the population of this pest and as many as 30 individuals of this tiny predator can complete their development on a single Asian long-horned beetle. At face value, introducing this small predatory beetle into areas where the Asian longhorn has become established seems like an

The Organisms That May Be Useful in Controlling the Asian Long-horned Beetle

Organism	Life cycle stage of beetle attacked	Potential as biocontrol agents
Entomopathogenic fungi		
Metarhizium anisopliae	Adults	Most virulent of the tested fungi
Beauveria bassiana	Adults	Similar virulence to *M. anisopliae*
B. brongniartii	Adults	Similar virulence to *M. anisopliae*
Isaria farinosa	Adults	Least virulent of the tested fungi
Nematodes		
Steinernema carpocapsae	Larvae	Potentially the most effective of the nematode species tested so far
Heterorhabditis bacteriophora	Larvae	Limited effectiveness
H. indica	Larvae	Limited effectiveness
H. marelatus	Larvae	The most effective of the *Heterorhabditis* species tested so far
Bacteria		
Bacillus thuringiensis	Larvae and adults	Not effective
Microsporidia		
Species not yet identified	Larvae	Infection prevalence in the wild is low
Parasitic wasps		
Scleroderma guani	Larvae	Found parasitizing 41.9–92.3% of larvae of a related European longhorn (*Saperda populnea*). An average of 45 adult wasps emerge from each host larvae.
Dolichomitus populneus	Larvae	Limited information on biology of this species
Predatory beetles		
Dastarcus longulus	Larvae and pupae	High prevalence of predation and many individuals develop on each host
Parasitic flies		
Billaea irrorata	Larvae	Attacks very early instars

excellent idea that should be implemented as soon as possible. Unfortunately, experience shows us that introducing a foreign organism to control another invasive species can be an ecological disaster (see cane toad entry). Without rigorous and exhaustive experiments we can never be sure how an exotic species will behave in a new environment. *Dastarcus longulus* could be introduced and make short work of the Asian long-horned beetle problem. Similarly, this small predator may behave completely differently in North America or Europe and instead of attacking the larvae and pupae of the Asian longhorn it may prefer indigenous wood-feeding insects, devastating their populations, disturbing forest ecosystems, and becoming a pest in its own right. This is one reason why biological control is such a complex and interesting field.

The simplest means of controlling and eradicating an infestation of the Asian long-horned beetle is identification and removal of infested trees, even those harboring very small numbers of larvae. Before an infested tree shows signs of stress, it is possible to identify the tell-tale signs of beetle attack. The small niches made by the females into which they lay their eggs can be seen on branches and trunks and the tunneling of first instar larvae is sometimes given away by sap runs and the presence of wood dust.

FURTHER READING

Food and Agriculture Organization of the United Nations [FAO]. *Global Review of Forest Pests and Diseases.* FAO Forestry Paper 156. FAO, Rome, 2009.

Hajek, A. E. Asian longhorned beetle: Ecology and control. In *Encyclopedia of Pest Management,* Vol. 2 (D. Pimentel, ed.). CRC Press, Boca Raton, LA, 2007.

Nowak, D. J., J. E. Pasek, R. A. Sequeira, D. E. Crane, and V. C. Mastro. Potential effect of *Anoplophora glabripennis* (Coleoptera: Cerambycidae) on urban trees in the United States. *J Econ Entomol* 94(2001): 116–22.

Smith, M. T., Z. Yang, F. Herard, R. Fuester, L. Bauer, L. Solter, M. Keena, and V. D'Amico. Biological control of *Anoplohora glabripennis* Motsch. A synthesis of current research programs. Proceedings of the USDA Interagency Research Forum—GTR-NE-300, 2003.

Bark Beetles

These tiny beetles are serious pests of forestry trees as well as ornamental trees. They are a problem around the world, but it is temperate regions that are most severely affected by their feeding and reproductive activities.

Brood gallery of a bark beetle. The vertical tunnel was excavated by the female and the horizontal channels are those excavated by her larvae. The small hole (labeled A) is the exit hole of a parasitoid wasp that fed on one of the bark beetle larva. (Courtesy of Ross Piper)

Taxonomically, the 6,000 or so known species of bark beetle used to be considered a separate family (scolytidea) within the order coleoptera, but relatively recent investigations of the DNA of these insects have revealed they are actually a very specialized group of weevils that have secondarily lost the distinctive snout that characterizes these beetles.

Bark beetles are considered pests for three main reasons. Their feeding activities can weaken trees and they serve as vehicles for a number of fungi species that are capable of devastating huge swaths of trees. The holes they make in the bark of the trees also allow other opportunistic pathogenic fungi and bacteria to invade and cause disease.

As their name suggests, the preferred microhabitat of the majority of bark beetle species for the purposes of reproduction is the bark of the host tree. The adults burrow their way through the outer bark. Within the deeper layers of the bark or directly beneath those layers, in the upper layers of the tree's sapwood, they excavate brood tunnels in which they lay their eggs. The reproductive biology of these beetles is very interesting as some species are monogamous, so a single male and female will construct a brood burrow, while other species are polygamous, that is, a male has a harem of several females, all of which construct their own brood burrows. When the larvae hatch they proceed to munch their own tunnels in the wood—tiny channels that snake away from the main brood tunnel. The patterns these brood and larval tunnels leave in the wood are distinctive enough to allow the identification of individual bark-beetle species even when there is no sign of the adults. Some bark beetle species tunnel into the inner wood of the tree rather than the phloem-bearing sapwood. These species can be considered a pest of timber rather than the tree itself as the tunnels they excavate reduce the quality of the wood for a number of purposes. In these species, the larvae feed on fungi, the spores of which are deposited by the female from a special pouch when she is laying her eggs. The fungal spores are carried by the larvae and sprout in the sheltered confines of the galleries. The introduced fungus feeds on the wood and in turn it is consumed by the beetle larvae. The bark beetles species that specialize in consuming the underside of the bark and the sapwood have little need for fungi as this material is easier to digest than the inner wood of the tree.

When small numbers of bark beetle are present, the damage they cause to a tree is minor, although they do permit the entry of fungi that can severely weaken and even kill the tree. However, in heavy infestations, the tunnels in the sapwood can be so numerous that the flow of sap through the trunk from the roots to the branches can be significantly impeded and the upper reaches of the tree do not receive the sap they need to sustain their biological processes. In these situations the tree may die.

The bark beetles have been living this way of life for many millions of years and in this time they have become acutely attuned to their hosts, sensing the best time to attack a tree, that is, when its defenses are at their weakest. Trees that have been damaged by storms or in forestry operations, diseased or moribund trees, and those experiencing drought or nutrient stress are very vulnerable to bark beetle attack. Trees experiencing stress for any cause will release compounds that the adult bark beetles are able to sense. The beetles home in on these chemical messages, enabling them to single out and attack the weakest trees. Once on or in the tree, the beetles are also known to produce aggregation pheromones that will attract other adult beetles, both males and females, to the breeding hot spot. These aggregation hormones are known to be triggered when the tree produces resins in response to insect damage. If large enough numbers of beetle are attracted, the infestation will overwhelm the tree's defenses.

In addition to the damage that bark beetles do simply by feeding on sap-conductive tissue, they can also cause severe damage to certain tree species because of the symbioses they have developed with certain species of fungi. A perfect example of this is the bark beetle *Scolytus multistriatus* and the fungus *Ophiostoma ulmi,* the causative agent of Dutch elm disease. Adult beetles of this species carry spores of this fungus and when they colonize a new host tree they inoculate the tree with the fungus, which eventually weakens and kills the tree, providing further future habitat for the beetles. In exchange for creating more habitat in which the beetles can breed, the fungus gets a free ride to hosts it might otherwise have no chance of reaching. There are countless bark beetles that have struck up symbioses with fungi, but the difference in this particular example is how devastating the fungi can be.

The origins of the fungus that causes Dutch elm disease are unknown, but it first appeared in Europe in the Netherlands in the early part of the 20th century. From the 1920s until the 1940s it ravaged the elm tree population of the northern hemisphere, eventually reaching North America in 1928. This original outbreak eventually disappeared, but in the 1960s an even more virulent form of the fungus (*Ophiostoma novo-ulmi*) emerged and killed off even more elm trees throughout Europe. In England alone, Dutch elm disease killed at least 74 percent of all the elm trees.

Not all the bark beetles are considered pests. It is only the small minority of species that attack tree species of commercial importance in large numbers, causing heavy infestations resulting in tree death over large areas of land. All types of trees are attacked by these beetles, but it is the commercial coniferous species that are most at risk because they are grown in monocultures that cover huge swaths of ground. The southern pine beetle (*Dendroctonus frontalis*) is a serious pest of pine trees in the southern U.S. states. It has been estimated that between 1960 and 1990, this species caused damage to pine forests worth around $900 million.

In Europe, the spruce bark beetle (*Ips typographus*) can cause very significant damage to forestry plantations where it attacks both damaged and healthy spruce trees. Compared to many bark beetles, *I. typographus* carries many types of pathogenic fungi that infect and kill host trees, such as *Ceratocystis polonica,* the bluestain fungus, which is highly virulent and also reduces the timber value of trees it doesn't kill by staining the wood blue (hence its common name). In Germany, a seven-year epidemic of this beetle that began at the end of World War II destroyed more than 30 million cubic meters of spruce. In the 1970s in Norway, an epidemic of this species destroyed 5 million cubic meters over an area of 140,000 square kilometers, which not only markedly changed the forest ecosystems by changing plant composition and reducing biodiversity but also had a significant effect on the country's gross national product. The same epidemic in Sweden killed around 7 million cubic meters of spruce. The spruce bark beetle is one species that could have a drastic effect on the pine forests of North America if it found its way there. In North America this beetle would undoubtedly kill huge numbers of trees and change the floral and faunal composition of wild and plantation forests alike. Needless to say, the U.S. authorities make every effort to ensure this pest never crosses the Atlantic.

The red turpentine beetle (*Dendroctonus valens*) is a pest of North American forestry, a native scolytid whose populations are regulated to a degree by the presence of natural enemies. At some point in the 1980s, this species was inadvertently transported to China, probably in forestry products, and since then it has gone on to spread through four Chinese provinces. In these provinces it has infested and killed more than six million pine trees (*Pinus tabulaeformis*) covering an area of half a million hectares of ecologically and economically valuable forest. The watersheds in these provinces are characterized by particularly thin soils easily washed away by heavy rains, and the forests fulfill an exceptionally important role in preventing the soil from being washed away into the many rivers of northern China. As an invasive species in China, the red turpentine beetle has few natural enemies, which exacerbates the problem.

There are numerous options for controlling bark beetles. Conceptually the simplest of these is imposing quarantine restrictions on the movement of timber and other forestry products, thereby preventing the spread of these insects into new areas. Although quarantine is theoretically straightforward, it is very complex and expensive in practice, especially in view of the huge borders that some countries have to police. As difficult as quarantine measures are to enforce, they can be very successful in keeping forestry pests out of a particular country.

Several chemical controls are also used to control bark beetles. Insecticides can be used on the small scale, but for large-scale control the difficulties in covering huge areas with these chemicals and the detrimental effect they have on the environment vastly outweigh the potential benefits. An area of bark beetle control research that is attracting a lot of attention is the use of semiochemicals, which are compounds that are identical to or that mimic those produced by insects and can be used to manipulate the behavior of the target species. The pheromones produced by bark beetles were mentioned above and it is these that scientists emulate to attract the beetles away from their host trees and into traps. Used appropriately, semiochemicals can be very successful in controlling bark beetle populations, even over huge swaths of ground. Biological agents can also be used successfully to control bark beetle populations. These agents include various fungi, nematodes, and other insects, all of which destroy bark beetles at various stages in their life cycle. In the case of the red turpentine beetle problem, a predatory beetle (*Rhizophagus grandis*) and a parasitic nematode

(*Steinernema ceratophorum*) have both been shown to control the populations of this pest.

Some bark beetle species can certainly be a nuisance for commercial forestry operations; however, the way in which forestry systems work, that is, the planting of huge areas of monoculture, is conducive to the survival of bark beetles species that use these trees as hosts. Maintaining biodiversity in these forests is one way of ensuring that bark beetle populations are regulated naturally and this means cultivating forests that reflect the composition of wild forests as much as possible.

FURTHER READING

Food and Agriculture Organization of the United Nations [FAO]. *Global Review of Forest Pests and Diseases.* FAO Forestry Paper 156. FAO, Rome, 2009.

Lieutier, F. Bark and Wood Boring Insects in Living Trees in Europe: A Synthesis. Springer, Amsterdam, 2004.

Bedbug

"Don't let the bedbugs bite." There are few people who can't have heard this familiar bedtime phrase and it's one that stems from a time when most dwellings were infested with this small bug. There are actually three species of bedbug that bite humans. *Cimex lectularius* is the most widespread and is the species most people will have encountered. *C. hemipterus* is found in tropical areas and *Leptocimex boueti* is restricted to West Africa. All three species have evolved from ancestors that fed on bats and cave-dwelling birds. Caves are excellent refuges and our ancestors probably started using them hundreds of thousands or millions of years ago. It was then that ancestors of the bedbugs started using our species as a source of food.

C. lectularius is the most important bedbug species from a pest perspective. We will only make reference to this species from here on, so when the term *bedbug* is used we mean *C. lectularius.* It seems this bedbug species originated in the Middle East, but its spread into Europe may not have occurred until historical times as the human race became increasingly mobile and trade prospered. By 400 B.C. this species had certainly reached Greece and in Italy the first references to it occur in 77 A.D. By the 11th century A.D., *C. lectularius* had reached Germany and by 1583 it

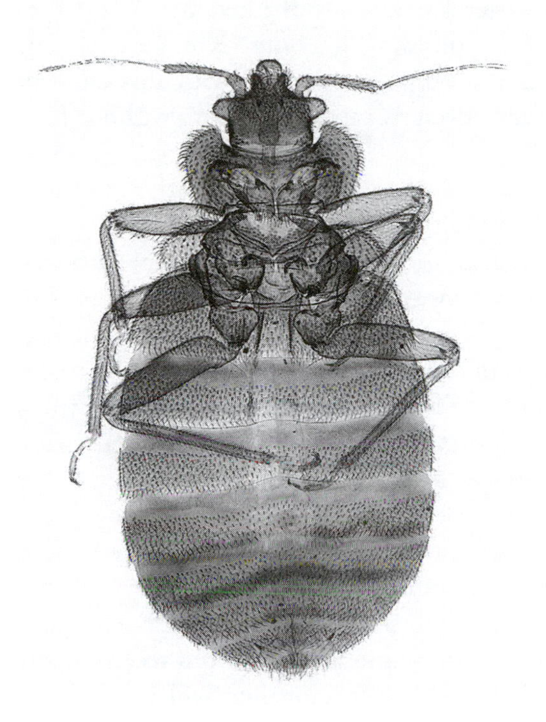

A bedbug at 40x magnification. (Warren Rosenberg | Dreamstime.com)

had arrived in England. Its arrival in England was at the same time as the wealthy courts of Europe were sending men and ships across the oceans to find new lands and the fabulous riches thereof. In the packing trunks and ruffed, period attire of these pioneers, *C. lectularius* probably found passage to every suitable corner of the globe.

Zoologically, the bedbug is a small, flattened, and wingless true bug. As adults they range in size from three to five millimeters and like other true bugs they have mouthparts adapted for piercing and sucking. They spend most of each day out of sight in tiny cracks and crevices, but as dawn approaches, when their human quarry is in a deep sleep, they emerge to feed. They locate a human from the carbon dioxide and heat we emit. When they've found a suitable place to feed, they plunge their sharp mouthparts into the skin. The bedbug injects an anesthetic so the victim is not disturbed as well as an anticoagulant to prevent the blood from clotting as it passes through the insect's narrow feeding tube. In a single feed, a bedbug can imbibe three times its own weight in blood. Once engorged, the swollen insect returns to its lair to digest its meal. Blood, although high in protein, is lacking in some other nutrients, so the bedbug has a pair of specialized organs, known as mycetomes, which house symbiotic bacteria responsible for supplementing the bug's blood diet. With this efficient digestion of its blood meals the bedbug is capable of going without food for at least 18 months.

The bedbug is not in the same class as insects such as the mosquito when it comes to being a public health menace simply because it doesn't transmit any disease-causing organisms. Twenty-seven species of pathogen have been shown to survive for varying lengths of time in bedbugs, but there is little evidence to suggest these insects ever act as vectors. The biggest problem with the bedbug is the annoyance factor. Around 90 percent of those people bitten develop some form of reaction to the bites, which can result in localized inflammation and itchiness. The author can attest to the irritation of these bites as one early morning he received nine bites in a line across his back. Closely spaced bites like this are common, as a feeding bedbug, if disturbed, will move off a little way and try to feed again. Not only do the bites cause intense irritation, but they can also become infected, which poses the risk of serious disease in areas with poor medical facilities. In cases of heavy infestations, victims can receive many hundreds of bites in the small hours of the morning, which is quite a significant blood loss, especially if it is maintained over many weeks or months.

Following the large-scale use of synthetic insecticides the bedbug was very nearly eradicated in many areas, but it is back with a vengeance, aided by insecticide resistance and globalization. More people are traveling to more destinations than ever before and the places they visit are areas where the bedbug has always been numerous. Thanks to their size and flattened shape these insects are expert stowaways. After feeding they will seek refuge in any suitable nook, even the seams of a large bag or jacket, and from there they are transported back to the traveler's home. They'll even jump ship on the public transport serving airports, train stations, and docks. In this way the bedbug is spreading like never before and as urban populations continue to rise populations of this insect will follow suit.

Controlling these insects is not as difficult as it may first seem. All that is required is a little diligence. After feeding, bedbugs will retire to any suitable crevice. This could be the tiny gaps in a bed frame, the gaps and holes in and around electrical and phone points, or a picture frame. The bugs betray their presence by small dark spots of feces around these hideaways. When an infestation is confirmed the affected rooms should be thoroughly

Bedbug Sex

They might be pests, but bedbugs are fascinating insects, especially when one considers the way in which they reproduce. These small bugs practice something called traumatic insemination. This is where the male pierces the body of the female and deposits his sperm in a specialized organ in the female's body cavity (the spermalege). From here, the sperm migrate through the female's body cavity to her reproductive tract, where they fertilize the eggs. This brutal means of reproduction carries quite a penalty for the female of the species: it can reduce her life span because the process causes mechanical damage and microorganisms can enter via the wound. How such a reproductive strategy evolved is something of an enigma, but it may be to overcome the problem presented by mate plugging. Many male animals, including many insects, try and guarantee paternity by blocking the female's reproductive tract with a gelatinous secretion following insemination. Traumatic insemination may be a means fertilizing a female's eggs even after the reproductive tract has been blocked in a previous mating. As this type of mating is detrimental to the health of the female, structures such as the spermalege have evolved as a way of reducing the detrimental impact of having the abdominal wall punctured.

Further Reading: Reinhardt, K., and M. T. Siva-Jothy. Biology of bed bugs (Cimicidae). *Annual Review of Entomology* 52(2007): 351–74.

cleaned and suitable insecticide powders or sprays applied over a period of weeks to kill the active bugs and any others that hatch from eggs. If there are no bites or evidence of feeding activity in the month following the last insecticide application, the bugs have been eradicated. Although eradicating bedbugs is relatively straightforward it can still be expensive, especially for the hospitality industry, as an infestation can sometimes involve huge complexes. For businesses such as hotels, the direct costs include the cost of treatment and cleaning, replacement of furniture, and compensation payments to guests. There is also the unquantifiable cost of a damaged reputation, even though an infestation of these bugs can occur in the most hygienic premises. Eradicating bedbugs may cost $75–150 per room, so a really big infestation requiring multiple insecticide treatments can cost thousands of dollars. It's not unheard of for large premises in urban areas to throw more than $100,000 at their bedbug problem. Pinning down the economic cost of the global bedbug problem is almost impossible, but it must be hundreds of millions of dollars every year.

FURTHER READING

Reinhardt, K., and M. T. Siva-Jothy. Biology of bed bugs (Cimicidae). *Annual Review of Entomology* 52(2007): 351–74.

Biting Midges

Biting midges are flies in the family ceratopogonidae. These insects have an ancient heritage: remains of specimens indistinguishable from modern species have been found in Lebanese amber, 120 million years old. These flies are so small that it's surprising they are capable of piercing the skin of large vertebrates at all, but bite they do and with a ferocity completely disproportionate to their diminutive dimensions. These biting flies are so well known and so widespread that almost everywhere in the world there are local names for these insects. In North America they are commonly known as no-see-ums or punkies, in Norway they are known as knotts, and in Polynesia they are called no-no's. They can be found from the

Biting midges are small insects equipped with sharp mouthparts for slashing the thick skin of vertebrates, including humans. They feed from the tiny pool of blood that forms. (CDC/ Dr. Richard Darsie)

tropics to the tundra of the Arctic Circle and depending on the species in question they can be found in enormous numbers. The family of flies to which the biting midges belong contains at least 6,000 species in 110 genera, but there are undoubtedly huge numbers of species in this family that are still unknown to science. Of the genera in this family, only four contain species that drink blood from mammals: *Culicoides* (the most important genus from a human and animal health perspective), *Forcipomyia, Austroconops,* and *Leptoconops.*

Biting midges are tiny insects. The adults are 1–2.5 millimeters long and they begin life as elongate eggs, no more than half a millimeter long. The females lay their eggs in a range of aquatic and semi-aquatic habitats, including swampy ground, salt marshes, tree holes, and animal dung. The larvae of many species of biting midge are very good swimmers capable of propelling themselves through the water with sinuous movements of their thin bodies. Exactly what the larvae feed on in the wild is very poorly known, but some are known to be predaceous on various tiny aquatic and semi-aquatic organisms, whereas others are herbivores or scavengers. It takes between two and seven days for the larvae to hatch from their eggs and depending on the species and where they live, larval development can take anywhere from two weeks to a whole year. A lengthy larval development period is especially characteristic of those species from high latitudes as they must cease their development during the winter and go into diapause to survive the cold conditions. It is not uncommon for some arctic species to require two years to complete their larval development. In these over-wintering species, the arrival of spring is the signal for them to continue their development into pupation and adulthood. Following the successful completion of pupation, the adults emerge and go about seeking mates and food.

Mating takes place when the midges form swarms above water or visual landmarks such as small bushes. These swarms are often composed of many species of midge and the males recognize the females of their species by the frequency of their wing-beat or species-specific pheromones. The females of many biting midge species mate only once; when they have obtained a batch of sperm they can go about finding food to complete the maturation of their eggs. However, there are some species where the females can save time in laying their first batch of eggs because they don't need a blood meal to complete the maturation of their eggs. Instead they depend on the energy reserves laid down during their time as larvae. Those species requiring a blood-fix to successfully mature their first load of eggs must seek out hosts. For some species, their choice of host is limited to one or a few species of related vertebrates, whereas many biting midges are generalists, happy to take blood from any vertebrate they can find. Amphibians, reptiles, birds, and mammals are all fair game for the biting midges. They locate hosts using visual and olfactory cues. In fair weather the females can disperse across five kilometers in 36 hours in their search for food. Male biting midges are not driven by this desire for blood, so they are much less vagile. Due to their very small size, the flight of biting midges is heavily dependent on the weather as wind-speeds of more than 2.5 meters per second can ground these flies.

Like the tabanids (see tabanids entry), these flies are pool feeders; they make a tiny slit in the host's skin and underlying capillaries with their mandibles and drink the blood and other fluids from the pool that forms. Like other blood-feeding flies, biting midges have massive appetites and an adult female *Culicoides* midge can drink her own body weight in blood during each meal. Following each meal, the midge retires to a suitably sheltered spot in nearby vegetation and begins the process of eliminating the excess fluids and salts from the food before digesting it.

Biting midges are universally loathed for two main reasons. Firstly, their bites are intensely irritating, as both the mechanical damage caused by the mandibles and also the saliva that spreads into the wound trigger an immediate inflammatory response that is localized but painful. The pain of a midge bite is something of a maladaptation when these flies are attempting to feed on humans as the host is quickly alerted to their presence. However, in areas where these flies abound, for every swatted midge many more get away fully engorged with blood. In high latitudes midges can occur in such profusion during the summer months as to render certain areas off limits to all but the most well-equipped or hardy visitors. The Highland midge (*Culicoides impunctatus*) occurs in mind-boggling densities, with 500,000 adults emerging from just two square meters of ground. Any large mammal venturing into these areas is descended upon by clouds of these flies—more than enough to prompt a hasty retreat. It has been estimated that losses to the Scottish tourism industry caused by this species amount to at least $440 million every year as visitors steer clear of the areas most affected. Not only is tourism affected by this species, but it is estimated that 20 percent of working days during the summer may be lost to outdoor jobs as even simple tasks are rendered impossible by the ravages of biting midges.

The nuisance aspect of biting midges is only one part of the story. These flies are also vectors for a number of pathogens, notably viruses, protozoa, and nematodes that can cause serious diseases in humans and livestock. The most important human pathogens transmitted by these flies are the virus that causes Oropouche fever and the nematodes that cause Mansonellosis. The former is not a life-threatening illness, but it can be debilitating, with typical fever symptoms lasting for a week or more. In the Amazon region of Brazil numerous outbreaks of Oropouche fever have been documented. Three species of nematode are responsible for causing Mansonellosis: *Mansonella ozzardi, M. perstans,* and *M. streptocerca.* These nematodes complete part of their life cycle in the body of the fly and infect humans to reproduce, but in most cases they do not cause any symptoms in their human host, although they can occasionally cause more serious problems such as blockage and inflammation of the lymphatic vessels and enlargement of the liver, especially when they occur in large numbers.

The most serious problem posed by the biting midges is their impact on livestock from the disease-causing pathogens they transmit, by far the most important of which is the virus responsible for bluetongue disease. Bluetongue disease was first described in South Africa in 1902 following the introduction there of European cattle and since then it has spread to Africa, Asia, South America, North America, the Middle East, India, and Australia. In very recent years (2001) its range has increased farther into the Mediterranean Basin and Europe. Transmitted by midges in the genus *Culicoides,* the disease afflicts a range of wild and domesticated ruminants, including cattle, sheep, goats, deer, and camelids, all of which display the typical symptoms of fever, excessive salivation, swelling of the mucus membranes in the nose and mouth, hemorrhaging from the mucus membranes of the mouth, erosion of mouth tissue, swelling and cyanosis of the tongue (causing the tongue to turn blue, hence the name of the disease), loss of wool, and depression and hemorrhages of the hooves resulting in lameness and difficulty standing. In cattle the disease is not often fatal, but in susceptible flocks of sheep mortality can be as high as 50–100 percent. Rather than mortality, the real problem with bluetongue is the extent of the breakouts and the impact on agricultural productivity, including weight loss of affected livestock, reduced milk yield, abortion, the veterinary costs of treating sick animals, and the loss of trade that results from the bans imposed on animal movements to contain the spread of breakouts.

In recent years, the breakouts of bluetongue disease have been exceptionally damaging to the livestock industry. For example, during the epidemic in Italy in 2000–2001, approximately 18 percent (263,000) of the nation's sheep and goat flock showed symptoms, and of these diseased animals, 48,000 died (3% of the nation's flock). The following year in the same country saw a similar proportion of animals infected, but mortality rose to 5 percent. As has already been mentioned, estimating the cost of an outbreak involves taking into account a number of factors, but the 2007 outbreak in France was estimated to have cost $1.4 billion, while the annual costs in the United States from trade losses and associated testing of cattle for bluetongue virus status has been estimated at $130 million. One worrying trend of the bluetongue disease phenomenon is the way that it has spread north in recent years; the evidence suggests the range expansion of the vector midges is made possible by global warming. As the climate continues to warm bluetongue disease will undoubtedly become ever more problematic.

Large-scale control of biting midges is practically impossible, given the geographic range of these insects and the enormity of their populations. Conventional strategies involve the use of insecticides against the larvae and adults, but for the reasons stated above these strategies generally have a very limited impact and the implications for the environment of large-scale applications of these synthetic chemicals can be very serious. The single most effective way of keeping these flies from biting humans and livestock is to avoid areas and times when the adults are on the wing and most likely to bite. Livestock can be moved to shelters during the evenings and early morning and humans can cover up, seek shelter, and make use of the numerous types of insect repellent on the market that deter the flies from landing and biting.

FURTHER READING

Blackwell, A. The Scottish biting midge, *Culicoides impunctatus* Goetghebuer: Current research status and prospects for future control. *Veterinary Bulletin* 71(2001): 2R–8R.

Boorman, J. *Biting midges* (Ceratopogonidae). In *Medical Insects and Arachnids* (P. Lane and R. W. Crosskey, eds.), pp. 288–309. Chapman & Hall, New York, 1993.

Purse, B. V., P. S. Mellor, D. J. Rogers, A. R. Samuel, P.P.C. Mertens, and M. Baylis. Climate change and the recent emergence of bluetongue in Europe. *Nature Reviews Microbiology* 3(2005): 171–81.

Black Flies

Black flies, also known as buffalo gnats because of their humpbacked appearance, are small insects, typically one to five millimeters long, belonging to the fly family simuliidae. Worldwide they are represented by around 1,500 species, of which several are important pests. A fossil pupa from the Jurassic period is virtually indistinguishable from that of some living species, suggesting these flies have been biting vertebrates for at least 100 million years. Black flies are found all over the world, but they are at their most abundant in northern temperate and subarctic zones where they can often appear in huge numbers.

As with many pest insects, the life history of these flies is very interesting. Their larvae are aquatic and they can only develop successfully in running, well-oxygenated water; therefore the females must deposit their 200–500 eggs on or in the water. Some species lay their eggs on the surface of water where they rapidly sink, while other species clamber down rocks or aquatic vegetation to lay their eggs directly underwater. The black fly larva is a fascinating little creature, capable of producing and using silk in an amazing way. The silk is produced by a pair of enormous glands that

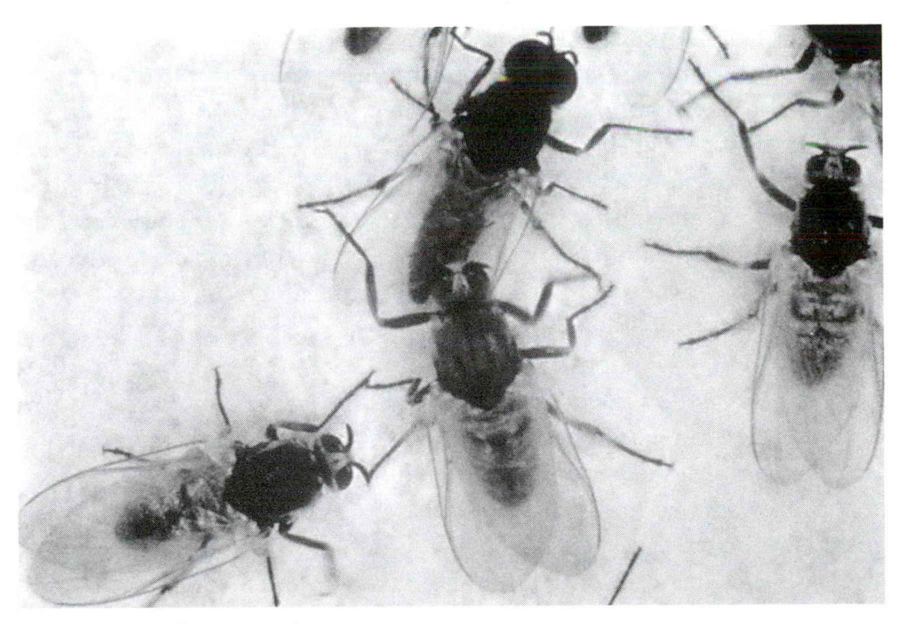

Black flies, like biting midges, feed on vertebrate blood by making a small incision in the skin. (World Health Organization)

stretch almost the entire length of the larva, issuing from pores in its head. As soon as the larva hatches it spins a silken mat on an underwater object, possibly a large stone, and clings onto the mat using the tiny hooked prolegs at the hind end of its body. Securely anchored against the powerful current the larva can commence feeding and it does this, primarily, by filtering the passing water for edible particles using its modified mouthparts, akin to a delicate pair of fans. Some species are even hitchhikers on other aquatic animals and will attach their silken pad to mayfly nymphs and freshwater crabs. Although the larvae are largely sedentary, they are capable of acrobatic movements. If disturbed or seeking new places to feed, they will stretch out, spin a new silken pad, grab onto it with their mandibles, and attach themselves using their prolegs. They are also able to move greater distances by releasing their grip on the silken pad and drifting downstream, often on a lifeline of silk, until they grip another submerged object and spin another pad. In suitable locations black fly larva can be present in incredible densities and the downstream side of a rock or log can be festooned with these little animals. Larval development depends on temperature, so some tropical species can go from egg to adult in four days, while species in subarctic areas can take half a year to complete their development. Once pupation is complete the adult flies extract oxygen from the water to fill internal air sacs, allowing them to shoot to the surface of the water like a cork. They then fly off to mate, and the females start seeking out large terrestrial animals from which to take blood meals in order to mature their eggs.

Black flies do transmit pathogens, namely the nematode worms that cause onchocerciasis, a nasty disease commonly known as river blindness. This disease is endemic to 34 countries in Africa, the Americas, and the Middle East, but sub-Saharan Africa is most severely affected. Globally, onchocerciasis is the second leading infectious cause of blindness, and although it's not fatal, it is debilitating for both individuals and whole communities. At the time of a World Health Organization review in 1993, more than 17 million people were thought to have the disease, of which 268,000 were blind as a result. Blindness is the most serious symptom of onchocerciasis and it is caused by the body's immune response to the juvenile nematodes that find their way to the eye, and/or to the bacteria living inside these nematodes. In addition to blindness the nematodes can also cause disfigurement, including lumps and severe swelling, dermatitis, and depigmentation of the skin. In addition, there is evidence that onchocerciasis may cause epilepsy. The nematodes are transmitted to humans in the bite of the black fly and the disease they cause is so feared in some places that fertile river valleys are often abandoned because of the large black fly populations. Huge strides have been made in the control of this disease, as the drug ivermectin can kill the juvenile and adult nematodes in their human host. In the long term the nematodes responsible for causing onchocerciasis will undoubtedly evolve resistance to ivermectin, so new drugs will be required. Both adult and larval black flies can be controlled, but this necessitates the use of toxic insecticides that kill nontarget organisms, affect human health in poorly understood ways, and are limited in the long term by the emergence of resistance. Preventing the flies from biting is the simplest, most effective way to disrupt the transmission of the disease; this prevention requires nothing more complex than basic education, insect repellents, insect nets, and suitable clothing, all of which are effective at keeping the flies at bay.

In suitable habitats black flies can occur in such huge numbers that their bites can be a painful nuisance for humans and potentially deadly for livestock. The bite of one black fly is not a cause of much concern, besides general irritation, local pain, and inflammation, but when there are thousands or hundreds of thousands of them the cumulative effects of their bites can be deadly, especially for large animals such as cattle and horses. In 1923, a species of black fly (*S. colombaschense*) found in central and southern Europe was responsible for the deaths of at least 16,000 livestock. In the mid- to late 1940s, another black fly species (*S. arcticum*) killed more than a thousand cattle annually in Canada. Some black fly species are known to have saliva that is toxic to cattle and it is this combined with a huge number of bites over a period of time that probably causes the deaths of livestock in areas where enormous numbers of black fly emerge in the summer months.

FURTHER READING

Adler, P. H., and J. W. McCreadie. Black flies (*Simuliidae*). In *Medical and Veterinary Entomology* (G. R. Mullen and L. A. Durden, eds.), pp. 183–98. Academic Press, San Diego, CA, 2009.

Boll Weevil

Cotton, a ubiquitous, endlessly useful fiber, begins life as a white fluffy material surrounding the seeds of plants in the genus *Gossypium*—the cotton plant. This white fluff is processed into the cotton we are familiar

The boll weevil can devastate cotton crops. (iStockPhoto)

with, a material with a multitude of uses even in the age of high-tech synthetic fibers. Like any other plant, cotton has its enemies and with the increasing cultivation of *Gossypium* species in expansive monoculture these herbivorous and pathogenic organisms have become a greater and greater problem.

One of the most serious pests of cotton is the boll weevil (*Anthonomus grandis*), a beetle whose small size belies the immense, some say legendary, social, economic, and agricultural impact it has had over the years. To many scholars, the boll weevil is a pest without equal, an insect that in many parts of the United States is considered to have had an influence on society, history, and culture second only to the American Civil War. Widely thought to be a native of Central America, the boll weevil steadily edged into the United States from Mexico. The first definite records are from Brownsville, Texas, in 1892.

The life cycle of this unassuming beetle begins when overwintered adults from the previous year emerge from the leaf litter and soil in which they have spent the coldest months. These overwintering sites are typically natural habitats just outside the crop, such as woods. Following their emergence from their long sleep the weevils feed for about a week, mainly on pollen, but also on flower buds (cotton squares) and the nascent fruits (bolls). With their appetites sated, the beetles move on to the important business of reproduction and in no time at all the mature females are ready to deposit their fertilized eggs on their host plant. The female uses her strong mandibles at the tip of her rostrum to nibble a hole at the base of her preferred oviposition site—the cotton square, although developing bolls are also used. Into this hole the female deposits an egg (one1 of around 200) and she seals the cavity made by her mandibles with a gelatinous secretion. After 2–5 days the larva hatches and commences feeding on the contents of the flower bud, eventually causing the entire structure to yellow and fall off the plant. The legless grub grows rapidly. Depending on the temperature, the weevil is ready to pupate after 7–14 days and it does this still safely concealed within the withering flower of its host plant. Reordering the larva's structures into those of the adult takes 4–6 days, after which time it is ready to chew its way out of its nursery. The newly emerged adult feeds on other flower buds for a few days before seeking a mate to complete the life cycle. In ideal conditions, the life cycle of this beetle can be completed in as few as 16 days, enabling as many as seven generations to be squeezed into a single year. With every mature female capable of knocking out 200 eggs it is not difficult to see how these little weevils can devastate entire cotton crops.

The vast majority of the damage caused by this species is solely a result of the feeding activity of the larvae, as they cause the flower buds and the small bolls to die and drop from the plant. In a heavy infestation of these beetles, the crop can appear deceptively healthy with abundant green foliage, but on closer inspection there will be very few mature bolls. From the late 19th century until the early decades of the 20th century the boll weevil was an extremely important pest in all the cotton-growing areas of the United States. Its spread from its native range was apparently very rapid and with each passing year the weevil horde inexorably expanded its range by 40–160 miles. Fully winged weevils can fly relatively large distances, and because of their small size, more passive dispersal by strong winds and storms is also possible. The rapid spread of the weevils was aided by their inadvertent transport in shipments of cotton bales around the country. By 1922, the boll weevil had reached the eastern seaboard of the United States, and 85 percent of the cotton-growing areas were infested (around 600,000 square kilometers). Alabama, an important cotton-producing state, was heavily affected by the boll weevil during the peak of the problem. In 1914, this state had produced around 1.7 million bales of cotton, a yield that had dropped to around half a million bales in 1917 following the colonization of the boll weevil. It has been estimated that throughout the United States, this insect has cost cotton growers at least $13 billion. Today it is still costing the industry in the order of $300 million every year in losses and the expense of prevention and control.

The wider problems often tagged onto the boll weevil infestation stem from the agricultural practices in the areas in which cotton was grown. A lucrative crop, cotton was the primary cash crop in many areas, underpinning entire rural economies. The damage wrought by the rapidly spreading weevils was so intense that farmers had little time to find and perfect the growing of other crops, so the economy of entire areas was devastated just when the financial markets were also about to deliver a hammer blow to the entire U.S. economy in the shape of the Great Depression. The widespread destruction of the nation's most important cash crop and the Depression conspired to create a period of almost unparalleled financial hardship for many people in the cotton-growing states. Bankruptcy was commonplace and much of the populace was forced to leave the homes they had known all their lives to search for work elsewhere. It wasn't until the 1940s that the situation began to improve. The immortalization of this humble beetle in countless phrases, rhymes, and songs serve as a testament to its influence in shaping the United States in the first half of the 20th century.

During the peak of the boll weevil problem, suggestions for ways to control and eradicate the insects were almost as numerous the weevils themselves. This was before the days of insecticides, so growers tried various methods to suppress the populations of the beetles, many of which were completely futile. Plants were doused with kerosene, ash, and anything that appeared to bring about the demise of the weevils without destroying the precious cotton plants. People went out into the field and collected the infested cotton squares by hand. The woodland and other natural and seminatural habitats surrounding cotton plantations were burned to deny the adult weevils the overwintering sites they required at the end of the growing season. The government established quarantine measures to limit the spread of the beetles, but most of these were next to useless. The only respite came in the 1930s when shifts in agricultural trends allowed weevil populations to be controlled. Farmers eventually adapted to grow other crops besides cotton, and with an increasing understanding of the beetle's biology they found that lower winter temperatures kept the weevil numbers somewhat at bay. The cessation of World War II was associated with a development that provided famers with another weapon against the weevils—synthetic insecticides, the commercial production of which was made possible by technological advances and the liberation of industrial manufacturing for purposes other than fabricating weapons of war. These first synthetic insecticides, such as DDT, were initially very successful. In no time at all the farmers became complacent and came to rely on these wonder chemicals, forgetting the cultural methods they had developed previously to suppress the weevil populations. Unfortunately, by the end of the 1950s, overuse of these synthetic insecticides caused natural selection of those weevils with chance genetic mutations, rendering them resistant to these chemicals. These weevils survived the chemical warfare and went on to spawn progeny similarly resistant to these insecticides. Before long, resistance was rife and these chemicals, once lauded as the nail in the coffin of this pesky insect, were becoming useless.

It was not long after that the devastating environmental effects of these first synthetic insecticides were appreciated—not a moment too soon for wildlife. New synthetic insecticides were developed and the pattern was exactly the same—exaggerated claims by manufacturers and users, overuse, evolution of resistance, and damaging environmental side effects.

Eventually it dawned on U.S. authorities that insecticides were not the solution to the boll weevil problem and in 1962 an entire laboratory was set up with the objective of exploring ways in which this beetle could be controlled and even eradicated. The scientists in this laboratory developed a whole raft of measures for controlling the weevil, collectively known as integrated pest management (see introduction). Over time, this has come to include traditional chemical control with insecticides, but in a very targeted fashion: pheromone-based traps, cultural practices, and the harnessing of natural enemies. To date, this integrated strategy has been very successful and the weevil has been eradicated from many areas where it once ran amok through the cotton crop.

The boll weevil problem is only a fraction of what it used to be, but this insect is still responsible for considerable crop losses in the United States and in other parts of the Americas, notably South America, where it has managed to invade the cotton-growing areas of Brazil, an area long thought to be out of reach of this pest due to a natural barrier—the vast swath of equatorial forest carpeting the Amazon basin. The beetle was first recorded in São Paulo State, Brazil, in 1983, perhaps introduced in a cotton shipment from infested areas. Since then it has gone on to infest the vast majority of cotton farms in Brazil and is widely regarded as the single most important pest of cotton in this country.

The negative aspects of the boll weevil's depredations are impossible to ignore, but this is just one side of the coin. The boll weevil problem in the United States stimulated the emergence of pest science as a discipline in its own right, a branch of biology responsible for huge leaps in agricultural productivity and a more efficient use of cultivated land. A fundamental element in the emergence of pest science as a rigorous discipline was recognition of the importance of intently studying the biology of pests and natural enemies. This basic, albeit hard-won, information was crucial in identifying those measures, which could reasonably be assumed to help suppress the populations of a pest.

The boll weevil story is a stark reminder of the follies of intensive agriculture and how a seemingly insignificant animal can affect not only farmers, but entire communities and regions.

FURTHER READING

Cross, W. H. Biology, control and eradication of the boll weevil. *Annual Review of Entomology* 18(1973): 17–46.

Haney, P. B., W. J. Lewis, and W. R. Lambert. *Cotton Production and the Boll Weevil in Georgia: History, Cost of Control and Benefits of Eradication.* Research Bulletin No. 428. Georgia Agricultural Experiment Stations, University of Georgia, Athens, 1996.

Hardee, D. D., and F. A. Harris. Eradicating the boll weevil. *American Entomologist* (2003): 62–97.

Smith, R. H. *History of the Boll Weevil in Alabama, 1910–2007.* Bulletin No. 670. Alabama Agricultural Experiment Station, University of Alabama at Auburn, 2007.

Citrus Leaf Miner

Leaf mining is a way of life peculiar to the insects, specifically the flies, beetles, moths, and sawflies. All the immature stages of a leaf mining species develop beneath the epidermis of the leaf in the tissue known as mesophyll, which contains the cells responsible for converting the sun's rays into chemical energy via the process of photosynthesis. The leaf miner consumes this photosynthetic tissue as it burrows through this impossibly thin layer, forming very obvious snaking tunnels or blotches in the leaf demarcated by the papery, translucent remnants of the epidermis.

This tiny moth is capable of causing severe damage in citrus crops. (Nigel Cattlin/ Visuals Unlimited, Inc.)

Just how this unusual life strategy came to evolve is not initially obvious. Why should a herbivorous insect go to the lengths of tunneling through the mesophyll of a leaf when it could more easily just remain on the outside and eat the leaf tissue in its entirety—epidermis and all? The answer probably relates to predation. Herbivorous insects, especially the immature stages, are at the mercy of a horde of predators, including other arthropods and vertebrates. Tunneling into the leaf gives these animals a physical albeit thin barrier between themselves and their many enemies. More importantly, many small predators and parasitoids locate their quarry by following the trail of odor that emanates from their feces and other waste. The frass of leaf mining insects is retained in the feeding tunnel, which helps to conceal the telltale odors from the herbivore's many enemies. The number of leaf mining species and the fact that this way of life has evolved independently in at least four insect orders attest to the success of this developmental strategy.

Numerous leaf miners are considered to be pests of agricultural and horticultural crops, but the species we are focusing on here is the citrus leaf miner (*Phyllocnistis citrella*), a tiny, cosmopolitan moth with a wingspan of around four millimeters that feeds on citrus and related plants. The gravid adult females deposit their eggs singly on the underside of the young leaves of the host plants and after 2–10 days the first instar larva hatches and bores through the epidermis of the leaf into the mesophyll. Once in the mesophyll, the larva munches its way through this tissue, leaving a serpentine tunnel. Normally, a single leaf supports only one larva of the citrus leaf miner, but in heavy infestations on large leaves there may be multiple snaking mines formed by as many as nine larvae. Depending on the temperature, the larvae complete their development in 5–20 days, progressing through four instars. With pupation imminent, the larva tunnels to the margin of the leaf that has nourished it and excavates a small chamber in which it will make the transition into adulthood, a process taking anywhere between 6 and 22 days. The short-lived adults emerge from their pupal cells during the dawn to seek members of the opposite sex to mate with. The females, replete with fertilized eggs, are active at dusk and during the night, searching for suitable sites to deposit their own eggs. The whole life cycle of this moth is completed in 13–52 days. In some areas, such as Northern India, the citrus leaf miner can go through as many as 13 generations in a single year, whereas 6 generations may be completed in a year in southern Japan.

Species such as the citrus leaf miner have evolved with flowering plants and in natural conditions the damage they cause is rarely significant; however, in monoculture cultivation their populations can grow to such a level they become problematic. The citrus leaf miner can be a problem for citrus growers, particularly in nurseries where young plants are more susceptible to being weakened by the damage caused by the insect's leaf tunneling. In heavy infestations with more than one larva per leaf, the tunneling reduces the ability of the leaf to photosynthesize effectively, which is exacerbated by leaf curling and withering. To date, this moth is known from most citrus-growing regions around the world, but the extent of the damage it causes and the economic losses have yet to be accurately quantified. Globally, around 70 million tonnes of citrus are produced every year, with the United States and countries in South America being among the largest producers. The potential of this moth to cause losses in this industry is very significant and growers around the world are investing time and money in monitoring the populations of this insect and the damage it causes.

Where infestations of the citrus leaf miner do occur, growers have various options at their disposal for controlling this moth. Insecticides can be used, but their environmental toxicity and limited effectiveness against a target safely concealed beneath the epidermis of the crop means they are far from the ideal choice. Other forms of chemical control, particularly pheromone-based traps, are good at reducing the moth population in a way that does not dent the populations of natural enemies. Additionally, pheromone traps can be used to monitor the size of an infestation in a particular growing area.

Although the citrus leaf miner is protected from many of its predators in the tunnels it excavates in the leaves of its host plant, it is not without its enemies. The most sophisticated and elegantly adapted of these are the various parasitoid wasps that deposit their eggs in or on the larvae and pupae of this tiny moth through the plant's epidermis. In Southeast Asia, Japan, and Australia the citrus leaf miner parasitoid fauna is impressive, currently standing at around 39 species, with many others undoubtedly to be identified. Wherever this moth is found, there will be at least one species of parasitoid wasp that attacks its immature stages. Managing the crop in such a way as to provide an environment conducive to the survival of these natural enemies is one of the basic tenets of biological control. It is also possible to augment the natural population of parasitoid wasps with commercially available individuals, providing enhanced control if an infestation develops.

FURTHER READING

Capinera, J. L. *Encyclopedia of Entomology,* Vol. 2. Springer, Dordrecht, Germany, 2008.

Hill, D. S. *The Economic Importance of Insects.* Chapman & Hall, London, 1997.

Pimental, D. *Encyclopedia of Pest Management.* CRC Press, Boca Raton, LA, 2002.

Cockroaches

Cockroaches are among the most well-known of all pests. The very word *cockroach* conjures up images of mess and filth, but as unpleasant as these insects may appear, they are very adaptable survivors that have simply taken advantage of the opportunities presented by human civilization. Cockroaches are among the most ancient insects, having roots somewhere in the steamy forests of the Carboniferous era, at least 300 million years ago. The general morphology and lifestyle of cockroaches has changed

A cockroach feeding on a pear. These adaptable and ancient insects are a fixture of the urban environment. (iStockPhoto)

little since they first evolved, attesting to the success of this original, rather primitive template. Today, at least 4,500 species of cockroach are known, although the actual number is probably far higher as the humid tropics undoubtedly support many species still unknown to science. Of all these species there are only around 17 that can be considered to be pests, and of these only about six species are significant (see sidebar).

The form and biology of cockroaches is primitive; the very first insects were probably very similar to modern cockroaches in appearance and lifestyle. These primitive characteristics include simple legs adapted for running, two pairs of very similar wings (in many species), nonspecialized, chewing mouthparts, and the absence of metamorphosis (cockroaches develop through several nymphal stages—essentially miniature adults). Cockroach biology has been very well studied because many species are seen as pests and they are also very easy to rear in captivity, making them ideal model organisms. From an ecological point of view they are typically nocturnal or crepuscular scavengers able to survive on all manner of food, including stored food, human and animal waste, and dead plants and animals. Therefore, in densely populated urban environments they are really in their element as food is in abundance and there are lots of nooks and crannies to hide in.

A cockroach's life begins as an egg. Depending on the species, a female can lay between 12 and 44 eggs, all of which are deposited in a special structure she secretes called the ootheca. This secretion hardens to form a leathery cocoon that protects the eggs from getting squashed and drying out. The females of some species carry this ootheca with them wherever they go, whereas others deposit the egg case in a safe place and have nothing more to do with their offspring. There are even cockroach species where the maternal female retracts the completed ootheca back into her body and the eggs develop and hatch in her genital pouch. Following the hatching of the eggs, 5 to 13 instars of nymphs ensue, depending on the species, living much as the adults do: skulking around in the shadows eating just about anything. Cockroaches are gregarious animals and it is normal to encounter them in large groups consisting of adults and nymphs at varying stages of development. Compared with many insects, cockroaches are long-lived and the individuals of certain species can live for well over a year even without food or water. Female American cockroaches can survive for 42 days without food or water and virgin female *Eublaberus posticus* can, remarkably, survive for 360 days on just water. The longevity of cockroaches and their ability to survive on very thin pickings are two of the most important factors in their global success, a success that puts them directly at odds with humans.

Of all the animals that live in association with humans, cockroaches are perhaps the most intimate of all these unwanted guests in that they live their whole lives in very close proximity to us. Throughout the world, cockroaches occur in large numbers, infesting countless homes and workplaces. For example, it is estimated that at least 24.5% of U.S. households use treatments designed to kill these insects on a regular basis—that's more than 20 million homes—making cockroaches the number one household pest in the United States. Most other animal pests of homes and workplaces often only make fleeting visits to find food, but the cockroaches call our houses, offices, and warehouses home, much to the annoyance of the two-legged occupants. Our problem with cockroaches stems from that fact they can negatively impact human and domesticated animal health via the transmission of pathogens as well as eliciting immune responses in sensitive individuals. Not only can they be detrimental to human and animal health, but they also consume and contaminate food intended for humans, pets, and livestock, which in serious infestations can amount to very significant economic losses.

One putative problem with cockroaches is bites, which are commonly reported, especially in heavily infested homes, but it is not known if these insects are genuinely responsible. The hands, feet, and faces of sleeping humans bear the wounds of apparent nocturnal, cockroach nibbling. The thought of cockroaches scurrying across their faces at night is one reason why many people have an irrational fear of these insects, but as we'll see these fears are not completely unfounded.

Like the housefly, cockroaches are magnets for pathogens because their bodies are clothed in scales and bristles that collect matter as they're scuttling around in filth. At least 32 species of bacteria have been isolated from cockroaches, including *Bacilus subtilis, Escherichia coli, Salmonella* species, and several *Proteus* species, which are responsible for diseases such as conjunctivitis, food poisoning, gastroenteritis, and skin and soft tissue infections. The unpleasant passengers of cockroaches also include fungi, protozoa, viruses, and the eggs of parasitic worms. In many situations these pathogens may not be problem, but when cockroaches are wandering surfaces in kitchens and hospitals it's easy to see how they disseminate disease far and wide. There is also some evidence that cockroaches may act as intermediate hosts for a number of parasites of humans, pets, and livestock. The infective stages of these parasites may find their way into the definitive hosts, us and our animals, via cockroach feces or fluids from their crushed bodies. Aside from their ability to spread disease, cockroaches are also infamous for causing allergic reactions in sensitive people. These reactions are typically caused by fragments of cockroach cuticle and feces that get into the body through the lungs or minute wounds in the skin. The typical allergic symptoms include eye and nose irritation and difficulty breathing in severe cases, which may even progress to anaphylaxis in hypersensitive individuals.

Although cockroaches are a problem the world over, they can be relatively easy to control and there are a number of common-sense solutions for eliminating them from houses and other places where people and domesticated animals spend a lot of time. Firstly and most simply is good hygiene as they quite happily eat any food or food debris left lying around and they will find lots of places to hide in accumulated rubbish. Therefore, cleaning up can deny cockroaches food and refuges. Traps of varying designs and vacuum cleaners can reduce the number of cockroaches in any given location. There is also the possibility of employing the service of biological control agents as there are a number of parasitoid wasps and parasitic fungi and nematodes that attack and kill cockroaches. These organisms are supplied by specialist companies and they can be released in problem areas to control the cockroach numbers in combination with improved hygiene and physical means of control. Although these techniques can eliminate a

The Important Cockroach Pest Species, Their Origins, and Their Current Geographic Distribution

Species	Geographic origin	Current geographic distribution
German cockroach (*Blatella germanica*)	Northern or eastern Africa or Asia	Worldwide
Brownbanded cockroach (*Supella longipalpa*)	Tropical Africa	Origin and North America and Europe
Oriental cockroach (*Blatta orientalis*)	Northern Africa	Origin and Europe, the Americas, western Asia
Smoky brown cockroach (*Periplaneta fuliginosa*)	Tropical Africa	Origin and North America, but more abundant in the Southern states
American cockroach (*Periplaneta americana*)	Tropical Africa	Worldwide
Turkestan cockroach (*Blatta lateralis*)	North Africa, Middle East, Central Asia	Origin and southern California to Texas

cockroach problem there are certain circumstances where chemical control may be the only option, in which case a large number of insecticides are available in formulations specifically designed for safe use around the home so that children and pets are not at risk of accidental ingestion. There are even sophisticated insecticides available, known as insect growth regulators, which interfere with the development of the cockroach, resulting in the death of nymphs as they shed their skin in order to grow. These are very specific to insects and their toxicity to mammals is very minimal.

Regardless of all the ways in which cockroaches can be controlled, these insects will never be eradicated because they thrive in the situations created by humans. Like the rat and housefly, those other great opportunists, cockroaches are simply making a living wherever they can, so from a purely zoological standpoint they are remarkably successful animals that have spread around the globe feeding on what we leave behind.

FURTHER READING

Bell, W. J., L. M. Roth, and C. A. Nalepa. *Cockroaches: Ecology, Behavior, and Natural History.* John Hopkins University Press, Baltimore, MD, 2007.

Rust, M. K., J. M. Owens, and D. A. Reierson. *Understanding and Controlling the German Cockroach.* Oxford University Press, New York, 1995.

Colorado Potato Beetle

This colorful beetle (*Leptinotarsa decemlineata*) exemplifies perfectly how human modification of the environment can result in a seemingly benign species becoming a pest. This large chrysomelid beetle is a native of the southwestern United States and Mexico. It first became known to science in 1824 following the collection of specimens some 13 years before. Before the arrival of Europeans and their crops in the Americas this beetle fed on plants such as buffalobur, a member of the genus *Solanum,* the same group of plants to which the potato and tomato belong. Seemingly harmless, this chrysomelid was merely another handsome beetle and it attracted little attention.

This all changed when settlers of European descent started farming in the beetle's native range. Among the plants they tended was the humble potato, itself a native of the Americas and one that had made quite a

Eggs, larva, and adult of the Colorado Potato beetle, one of the most destructive of all insect pests. (Jeff Daly / Visuals Unlimited, Inc.)

circuitous route to get to the soils of the southwestern United States. It had originated in the Andes and was among some of the treasures the first European explorers to the New World returned home with more than 300 years previously. In those intervening years, plant breeders and farmers had gone to work on this interesting plant and modified some of its characteristics to develop varieties capable of producing bounteous crops of nutritious food in the temperate Old World. This selective breeding had gone on in the distant fields of Europe, during which time the potatoes were free of their natural enemies, including insects. By the time potatoes returned to the New World they had probably lost the edge that helped them fight off the attacks of plant-feeding insects. The black and yellow chrysomelid soon discovered the new, improved, palatable potato and it rapidly switched host plants, forsaking the wild, well-defended *Solanums* of its native range. The first recorded major outbreak of what was soon to be christened the Colorado potato beetle was in 1859 in the fields of Nebraska and this insect has been a serious pest of potatoes ever since, with no sign of giving up anytime soon.

From this earliest recorded outbreak the Colorado potato beetle spread fast and it had reached the Atlantic coast of the United States and Canada before 1880, which is a distance of at least 1,800 kilometers. This means the beetles advanced at a rate of around 80 kilometers a year. Many may have achieved this under their own steam, as the adult beetles are strong flyers, but it is also likely many were inadvertently moved around the country in potato shipments; the species pupates and also overwinters as an adult in the soil, so potatoes in transit surrounded by the earth they were dug up from may have harbored the insect. The beetle is also a pest of eggplants and tomatoes, so the movement of any of these crops could have hastened its spread. Not too long after their arrival at the east coast of North America the Colorado potato beetle made its first appearance in Europe (France, to be exact, in 1922), and by the end of the 20th century it had made its presence known throughout Europe and into Asia Minor, Iran, Central Asia, and western China. Today, its range covers about 16 million square kilometers on two continents and the worrying thing is that it continues to spread. Potentially, the beetle could survive in any location where potatoes can be grown, including temperate areas of East Asia, the Indian subcontinent, South America, Africa, New Zealand, and Australia. Areas that are currently free of this beetle do their utmost to try and keep it that way. Thanks to the English Channel and a strict policy for maintaining its Colorado beetle–free status, the United Kingdom has so far managed to keep this insect out, as have Ireland, Iceland, and the Scandinavian countries.

Many experts regard the Colorado potato beetle as something of a superpest. Once it made the initial host-plant switch back in the mid-19th century, its natural history assured its pest status. Both the adults and the larvae of this beetle are voracious plant-devouring machines, with the former being able to consume around 10 square centimeters of potato leaf a day, while the latter munch their way through around 40 square centimeters of potato leaf before they pupate. When thousands of these insects throng a potato field, it is easy to see how they can ruin entire crops. They are also able to produce large numbers of young, with a single female producing 300–800 eggs. Furthermore, they have no natural enemies in most of their present range, so mortality of the vulnerable eggs and larvae is very low. As mentioned above, the adults are strong fliers, presumably a trait that allowed them to find patches of their natural host plants in a florally diverse landscape. Voracious appetites, high fecundity, and excellent dispersal ability all combine to produce the perfect pest. Another trait that catapulted this beetle into the pest hall of fame is their incredible ability to render toxins harmless, which was only observed when agricultural intensification dictated the liberal application of various pesticides to control troublesome insects. This trait may have evolved as a way to make a living on host plants that are stuffed full with toxins to deter and even kill herbivorous animals. It's well known the wild *Solanums* are a very well-defended bunch of plants and to survive on this poisonous diet, the Colorado potato beetle evolved enhanced enzymatic pathways to neutralize these toxins. The beetles were even able to adapt the plant's defenses to their own ends by incorporating them into their own armory, thereby keeping their own enemies at bay. The Colorado potato beetle broadcasts its toxicity to its enemies with bold black and yellow markings (aposematism).

Globally, potatoes are the fifth most important food crop. In 2007, the entire world production of this tuber stood at 325 million tonnes, with China producing more than 20 percent of this total. In many parts of the world, particularly the more affluent countries, potatoes are a staple crop and on average, every human consumes 31 kilograms of potatoes per year. In some areas, the Colorado potato beetle can reduce yields by as much as 30 percent. Therefore it comes as no surprise that the Colorado potato beetle has long been the target of pesticide industry research in the hope of finding a means of control and even eradication. Hundreds of compounds have been tested against it and in many ways this struggle has molded the modern pesticide industry. Of the huge range of insecticides tested against it, none have been successful in denting the numbers or the spread of this pest. To date the Colorado potato beetle is resistant to 52 different compounds in all the major insecticide classes. Relying on just chemicals to control this or any other pest is doomed to failure (read more about this in the introduction). The only hope of controlling this beetle is by using a range of means (integrated pest management—see introduction), including potato husbandry, plants selectively bred or genetically engineered (see sidebar in the European corn borer entry) to be beetle resistant, and the harnessing of natural enemies.

Because Colorado potato beetles are capable of severely damaging crops of this important vegetable, attempts have been made to use them in biological warfare. During the Second World War the Germans misinterpreted intelligence from France and the United Kingdom and believed the allies were planning to disrupt the production of potatoes in Germany by releasing large numbers of this beetle. As a result, the Germans began their own research into using this insect in the war effort. The Germans realized the south coast of England supported about 400,000 hectares of potato fields, making this area the prime target for their efforts. There are anecdotal reports that beetle bombs were dropped on the Isle of Wight, but with little or no success. Following the Second World War, U.S. forces were accused of dropping Colorado potato beetles from planes flying over East Germany, accusations the United States flatly denied.

FURTHER READING

Alyokhin, A. Colorado potato beetle management on potatoes: Current challenges and future prospects. *Fruit, Vegetable and Cereal Science and Biotechnology* 3(Special Issue 1)(2009): 10–19.

European Corn Borer

This dowdy little moth is another example of the devastating impact of accidental animal introductions. As its name suggests, this insect is a native of Europe and in their native range the caterpillars of this moth can be found tunneling in the stems of many types of plant. At some point in the early 20th century this moth was somehow introduced to the Americas,

The life cycle of the European corn borer, a moth that can devastate corn crops. (National Geographic Society/Corbis)

first being recorded near Boston in Massachusetts in 1917. It has since become apparent there was more than one introduction and the founding moths came from more than one place in Europe. Almost a century has elapsed since the initial introduction of this insect to North America and in that time it has spread westward to the Rockies in both the United States and Canada and southward to the Gulf Coast states.

This species was an occasional pest in Europe, but it really came into its own when it reached stateside as the abundance of corn (*Zea mays*) was very much to its liking. Today, the annual worldwide production of this plant is in the order of 800 million tonnes, of which around 40 percent is produced in the United States; a considerable contribution to the agricultural might of this nation.

The caterpillars of this moth are such a problem for corn growers because they feed on the tassels, whorl, and leaf sheath tissue; they tunnel into the leaf midribs and eat pollen that collects behind the leaf sheath. They also feed on the silk, kernels, and cobs as well as tunneling into the plant's stem. It's typically the older caterpillars that tunnel into the plant and many consider this behavior to be the greatest cause of damage to the crop because when the movement of water and nutrients to the growing fruits is disrupted, the plant is structurally weakened. Harvesting becomes more difficult, and the numerous tunnels allow the entry of plant pathogens that can further weaken or even kill the plant and the developing cob. The voracious feeding activity of huge numbers of corn borer caterpillars can have a huge impact on the corn harvest throughout the corn-growing areas of the United States and Canada. This unassuming animal is the most damaging insect pest of corn in these countries. Quantifying the economic losses caused by the greedy chomping of these insects is impossible to assess, but estimates are in the order of one billion dollars every year, which includes crop losses and the expense of control measures. A particularly heavy outbreak in Minnesota in 1995 resulted in crop losses amounting to $285 million.

The predilection of these moths for corn is the reason why they're considered to be such a pest; however, the caterpillars are not fussy when it comes to food and they quite happily feed on a range of crops, including, the fruit and stems of beans, pepper, and cowpea, the stems of celery, potato, rhubarb, Swiss chard, and tomato, as well as the leaves of beet, spinach, and rhubarb. The catholic tastes of the corn borer and its distribution catapult it to the rank of superpest.

The economic losses caused by the European corn borer have stimulated the development of a whole avenue of scientific endeavor aimed solely at seeking a way to bring about the demise of this insect. However, try as we might to exterminate this moth we are certainly not going to see its departure from the cornfields of North America anytime soon. Like all crop pests, modern agriculture has produced a habitat in the shape of monocultures that suits species like the corn borer perfectly (see introduction). With agricultural practices unlikely to change anytime soon, the best we can hope for with this moth is control and to this end several techniques are at the corn grower's disposal.

Monitoring of the corn borer population tells the famer when numbers of the pest have risen to a level at which control measures would be economically viable. Monitoring can be carried out with light traps and traps baited with pheromones. Such monitoring can also be used to assess where the moth population, as a whole, is in the life cycle, which in turn dictates whether control measures should be brought to bear.

As with all insect pests, the old faithful when it comes to control of the corn borer is the application of insecticides. These have to be applied judiciously to avoid wastage and missing the caterpillars when they are at their most vulnerable. Unfortunately, this measure is flawed by the development of insecticide resistance, the increasingly important phenomenon that limits the effectiveness of all applications of these chemicals.

If farmers were to rely solely on insecticides for the control of European corn borer they'd quickly find themselves in trouble. Typically, chemicals are allied with other techniques to keep the moth populations controlled and within economic thresholds. Caterpillars from the second generation of moths in any given year like to overwinter in the stalks of the corn that remain in the ground after the harvest. Therefore, the stalks can be

Genetically Modified Crops and Pest Control

Genetic engineering allows the manipulation of an organism's genes and the insertion of genes from one organism into the genome of an unrelated organism. These techniques have the potential to revolutionize agriculture as scientists can select the traits that enhance crop production, by improving crop growth or the nutritional content of the crop and/or by minimizing the damage caused by pests and diseases. Many genetically modified (transgenic) plants currently exist, but there is significant opposition to the large-scale release of these plants from environmental organizations.

Some of these transgenic plants have been modified by inserting the genes carrying the instructions for producing bacterial toxins, an example of which is Bt maize. In this case the genes in question are from the bacteria, *Bacillus thuringiensis,* and as the plant grows it produces this toxin, serving as defense against the caterpillars of the corn borer moth (*Ostrinia nubilalis*). To date, this research has not proved to be very successful in terms of preventing damage caused by the corn borer.

The opposition to transgenic organisms is rooted in concerns that we can't predict how these organisms will behave in the wild. In the case of genetic engineering to prevent pest damage it's possible the engineered traits could negatively affect the populations of nontarget organisms. Plants can also be engineered to be resistant to herbicides, allowing the use of potent chemicals to control weeds without damaging the food crop. Opponents argue that if these herbicide resistance genes somehow found their way into weed species, superweeds could inadvertently be created.

Further Reading: Mchughen, A. *Pandora's Picnic Basket: The Potential and Hazards of Genetically Modified Foods.* Oxford University Press, Oxford, United Kingdom, 2000.

mown followed by plowing, which is effective at ending the lives of the caterpillars overwintering in the remnants of the crop. In northerly latitudes where the corn borer population has only one generation per year, late sowing of the corn crop can be effective as the shorter plants are less attractive to the egg-laying female moths.

Corn growers have another trick up their sleeve in the form of biological control. A host of different organisms, including bacteria and parasitic wasps, have been identified that infect or parasitize this pest. The bacteria, *Bacillus thuringiensis,* available in various preparations (e.g., liquid sprays) can be as effective as many chemical insecticides in some situations, although the use of this biological measure is typically hit or miss. Various parasitoid wasps, bred for the job, can be released into a corn crop to feed on the eggs and/or caterpillars of the moth. Again, in some circumstances this approach can be successful, but often only moderate levels of suppression are achieved.

In the age of biotechnology, cultivators of corn now have very sophisticated means of defeating the corn borer (see sidebar). Manipulation of DNA makes it possible to take the genes from one organism and insert them into the genome of another—hey presto—a living thing with the qualities of the scientist's choice. In this case, it's corn with a genome that contains the genes for producing a bacterial toxin. The plant tissues, suffused with the toxin, are greedily eaten by the caterpillars and the end result is a lot of dead caterpillars. In addition, advances in the understanding of genetics and plant breeding have also allowed the continued development of plants with resistance to these marauding insects.

FURTHER READING

Youngman, R. R., and E. R. Day. *European Corn Borer Fact Sheet.* Virginia Cooperative Extension Service, Virginia Tech, Blacksburg, VA, 1992.

Fleas

Fleas are wonderfully adapted parasites, primarily of mammals, but also of birds. Their adaptations to a parasitic way of life are among the most sophisticated of all the insects, with a number of unique characteristics that allow them to get on, stick to, and drink the blood of vertebrates. Their most remarkable feature is a pair of huge back legs, powered by the elastic properties of a protein known as resilin, which allows feats of jumping with few parallels in the animal kingdom. If the jump of a flea were scaled up to human dimensions it would be equivalent to you or I clearing a building more than 240 meters tall. The jump of the flea has evolved as a means of getting onto large animals from ground level in the absence of wings. Fleas have secondarily lost their wings because these delicate structures would quickly get damaged and torn as the insect negotiates the pelage of its host. Once it has managed to hop onto a host, the flea stays put thanks to the numerous spines and bristles adorning its body in strategic locations. Once securely in place the flea can pierce the host's skin with its sharp mouthparts to suck blood from the vessels and tissue beneath. The fleas share the ability of many wingless bloodsucking insects in that it can live without food for extended periods of time and certain species have been shown to survive periods of starvation lasting 125 days.

Fleas have a larval stage, which has important consequences for the types of host they can parasitize successfully. In the vast majority of flea

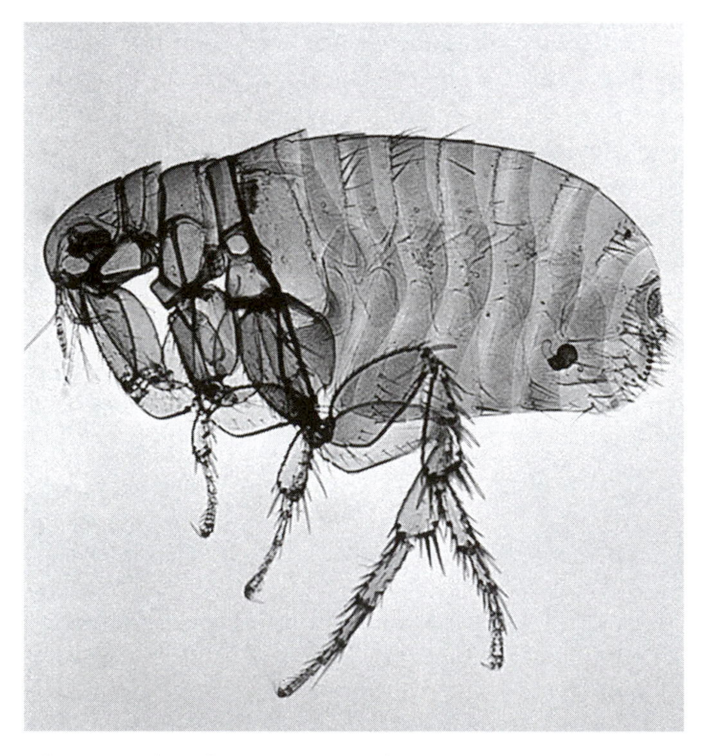

The oriental rat flea, the vector of the bacterium that causes bubonic plague. (CDC/ World Health Organization)

species, the larvae have no special adaptations for reaching or clinging onto their host; therefore they are nest dwellers and are limited to the area around the host where they feed on nest detritus and the excretions produced by the bloodsucking adults above. Animals that don't build nests or that don't return to the same hideaway to sleep cannot be used by fleas, perfect examples of which are the great apes. None of these primates build permanent nests, so none of them are troubled by any flea species. We humans, on the other hand, the most intelligent and settled of the primates, have houses and other dwellings, so we have our very own flea species—the human flea (*Pulex irritans*), although this parasite quite happily feeds on animals diverse as pigs, dogs, ground squirrels, and burrowing owls. The origins of this flea species are unclear, but it has closely related species in Central and South America and one theory is that the ancestors of this species were parasites of animals domesticated by Amerindians, such as guinea pigs or peccaries.

Well-adapted parasites they may be, but fleas have long been the target of human animosity. Their position in the collective human psyche is reflected in the English language and commonly used phrases such *fleabag, fleapit, flea market,* and *flea in the ear,* all of which imply shabby things or places or the irritation caused by these insects. More than 2,500 species of flea are known, only a few of which are considered pests based on their propensity for biting humans and domesticated animals, not to mention their role in transmitting disease. All fleas bite, which in itself is painless, but inflammation can occur after the flea has finished feeding and because the skin has been broken there is the potential for secondary infections to take hold, especially if the bite is scratched in order to ease the itching. These secondary bacterial infections can be life-threatening in areas where basic medical facilities are limited. Certain individuals can also become sensitized to flea bites, which can eventually lead to potentially serious allergies.

Flea bites are a nuisance, as any cat owner will tell you, but these insects are more of a health concern for the diseases they transmit. The ability of a flea species to act as a vector is dictated by the time it spends on a host. These parasites can be broadly grouped into three categories based on the amount of time they spend on one host:

1. Nest species—seldom on the host apart from feeding, but are abundant in the host's nest. Examples: fleas in the genera *Conorhinopsylla* and *Megarthroglossus*.
2. Mobile species—mostly found on the host and can easily move between host individuals. Example: the Oriental rat flea (*Xenopsylla cheopis*).

3. Sedentary species—found on the host and do not move between host individuals. Examples: the sticktight flea (*Echidnophaga gallinacean*) and the chigoe (*Tunga penetrans*).

The most important vector fleas are the mobile species as they can transmit bloodborne pathogens from one host to another. The most important mobile flea species from a public health perspective is the Oriental rat flea because it is the main vector of bubonic plague (see brown rat entry), a disease that killed millions of people throughout the 19th and 20th centuries and which is still killing people in developing countries today. Murine typhus, a bacterial disease caused by *Rickettsia mooseri* and *R. prowazekii* that manifests as a mild fever in humans, is also transmitted by the Oriental rat flea. In addition to its ability to move between hosts, another characteristic that makes a particular flea species an efficient disease vector is the ease with which its gut can become blocked by the growth of a pathogen, such as the bubonic plague bacterium (*Yersinia pestis*). The Oriental rat flea's narrow gut is rapidly blocked by bacteria and a new blood meal cannot pass; therefore the blood is regurgitated back into the host tainted with bacteria. Apart from these human diseases, fleas can also transmit other disease-causing organisms to other animals, including various viruses, protozoa, nematodes, and platyhelminthes, many of which can debilitate and kill pets and livestock.

Fleas are routinely controlled with insecticides applied to both the hosts and the larval habitat. In the home, the cat flea is a common problem and this species can be controlled by the application of insecticide directly to the host as well as carpets, furnishings, and so on that may harbor the eggs and larvae. Good hygiene is another way of controlling fleas in domestic settings as thorough cleaning destroys eggs, larvae, and adults. As with any insect pest, fleas can rapidly evolve resistance to insecticides, so these chemicals need to be used judiciously.

FURTHER READING

Buckland, P. C., and J. P. Sadler. A biogeography of the human flea, *Pulex irritans L.* (Siphonaptera: Pulicidae). *Journal of Biogeography* 16(Sup. 2)(1989): 115–20.

Durden, L. A., and N. C. Hinkle. Fleas *(Siphonaptera).* In *Medical and Veterinary Entomology* (G. R. Mullen and L. A. Durden, eds.), pp. 110–31. Academic Press, San Diego, CA, 2009.

Krasnov, B. R. *Functional and Evolutionary Ecology of Fleas: A Model for Ecological Parasitology.* Cambridge University Press, Cambridge, MA, 2008.

Gypsy Moth

The gypsy moth, *Lymantria dispar,* is one of the most important forestry pests in the world. A native of Europe, North Africa, Asia, and Japan, the gypsy moth was purposefully introduced to North America in 1868 or 1869 by the enterprising Frenchman, Etienne Leopold Trouvelot, an artist by profession who also had a passing interest in entomology. A resident of Medford, near Boston, Trouvelot became interested in using native North American silkworms for commercial silk production. During a visit to his homeland he decided to bring some gypsy moth eggs back to the United States with the intention of hybridizing the native American silkworms with the Old World gypsy moth to produce an insect capable of producing prodigious quantities of high-quality silk. Unfortunately, his plan was utter nonsense and no sooner had he installed the hatched gypsy moth caterpillars on trees in his back garden than some made good their escape to set up home in the Americas. Needless to say, Trouvelot soon lost interest in entomology and instead turned his attentions to astronomy, a pastime that involved simply observing rather than very costly meddling.

Today, the gypsy moth is found throughout much of eastern North America, with the largest populations in the eastern states of the United States. Isolated populations have been identified in western regions beyond the boundaries of the current continuous distribution, but to date these have been eradicated or have died out naturally. The gypsy moth only currently occupies about one-third of the North American landmass that it is potentially suited to, so there is a great deal at stake in checking the expansion of this insect and managing or even eradicating the established populations.

The life cycle of the gypsy moth begins with eggs deposited in egg cases by the females from the previous year. These hardy eggs are able to withstand the temperate winters. When spring arrives and leaves begin to unfurl, the caterpillars are ready to hatch and begin their assault on the greenery. After a few weeks of almost incessant feeding, the larvae are ready to pupate, which takes place on the host plant or in the ground. After one to two weeks, adults emerge to complete the life cycle. In most locations throughout their native range the female gypsy moths are flightless

The ravenous caterpillars of this moth can defoliate huge areas of forest in areas where they have been introduced. This female has just laid her eggs on the trunk of a host tree. (Bill Beatty/Visuals Unlimited, Inc.)

and it is up to the male to take to the air in order to find a mate. Not long after mating, the female lays her eggs and both she and her mate die, the purpose of their fleeting adult lives fulfilled.

It is the voracious caterpillars of the gypsy moth that make this animal a pest. Like all caterpillars, gypsy moth larvae are nothing more than eating machines able to plough through foliage at an astonishing rate. However, when they're small they are also capable of moving quite some distances, not by walking, but by taking to the air. Caterpillars don't have wings, but they are capable of producing silk and gypsy moth caterpillars can produce lots of this natural wonder substance. They secrete silk from spinnerets on their mouthparts and in order to take to the air they use a technique perfected by the spiders—ballooning. At the top of a bush or small tree, the newly hatched caterpillars each extrude a gossamer strand of silk that eventually grows long enough to cause drag sufficient to carry the tiny larvae aloft. In this way the caterpillars can be carried many meters or even more than a kilometer, enabling them to exploit new areas of habitat.

The dispersal ability of the gypsy moth caterpillars combined with the fact they have few natural enemies in the forests of North America means their populations can explode here. In severe outbreaks the caterpillars can completely defoliate many species of host plant, although their favored hosts are oaks (*Quercus* spp.) and aspen (*Populus* spp.). Even in severe outbreaks, a tree is able to tolerate heavy defoliation (61–100%) in one season as long as it isn't diseased or experiencing drought or nutrient stress. The problem lies with successive defoliations, which will kill even healthy trees. Since 1980, gypsy moth caterpillars have defoliated around 4,000 square kilometers of forest each year. The outbreak of 1981 was particularly severe with around 52,200 square kilometers of forest being defoliated. Such extensive damage has implications for entire forest ecosystems. Animals that depend on the host trees stripped by gypsy moth are left without food and the huge increase in the number of dead, standing trees can completely disrupt the delicate balance that exists among the many denizens of these forests. If the gypsy moth were allowed to spread unchecked throughout North America it could potentially change the entire landscape for the worse, denuding entire areas of their forests and reducing overall biodiversity. Financially, it has been estimated that the gypsy moth causes economic losses every year of around $30 million.

Addressing the gypsy moth problem is now far beyond eradication. The species has had almost 150 years to spread through a huge part of North America, so the best we can hope for is limiting further expansions of its range and regulating the populations in areas where it is well and truly established. Achieving these objectives will be expensive and logistically complex. Forest management can help to reduce the intensity of gypsy moth outbreaks and can also minimize economic losses following an outbreak. The moth's preferred host plants can be thinned and moribund tress following successive outbreaks can be removed to harvest the timber before the tree is completely dead.

Chemical control is still the main weapon against the gypsy moth, although in areas where it is already established, spraying toxic compounds over large areas is both economically unfeasible and environmentally unacceptable. Insecticides are mostly used to eradicate isolated populations beyond the current distribution in order to prevent the moth from spreading further across North America. Simple physical barriers around the bases of host trees and plants can prevent hungry, wandering caterpillars from climbing and attacking the foliage. Again, this can be successful on a small scale, but when we're talking about whole forests of millions of tress it's simply not feasible. Pheromone traps, based on the compounds produced by the female moths to attract mates are also used regularly for gypsy moth monitoring and control.

Biological control perhaps holds the most promise for controlling the gypsy moth. Various commercial preparations of insect pathogens, notably *Bacillus thuringiensis,* a virulent bacterium that infects and kills insect larvae, has proven its effectiveness in controlling gypsy moth populations. Similarly, there are countless predators and parasitoids of the gypsy moth in its native range. Several species of parasitoid wasp attack the gypsy moth and its very close relatives, making them suitable candidates for controlled releases in North America. These biological control agents provide us with a means of subjecting the gypsy moths in North America to the normal rigors of life this species faces in its native range. Throughout Eurasia outbreaks of this moth are not unheard of, but they are nowhere near as large or as destructive as those seen in North America because they are naturally regulated by pathogens, parasites, and predators.

FURTHER READING

Food and Agriculture Organization of the United Nations. *Global Review of Forest Pests and Diseases.* FAO Forestry Paper 156. FAO, Rome, 2009.

Head Louse and Body Louse

There are more than 5,000 species of louse and several of them are considered to be pests of humans or livestock (see sidebar). Two species feed directly on humans: *Pediculus humanus* (which is divided into two subspecies: the head louse, *P. humanus capitis,* and the body louse, *P. humanus humanus)* and the pubic louse (*Pthirus pubis*) (see pubic louse entry). The evolutionary trajectories of primates and lice have been entwined for at least 25 million years and as the primate line diverged into the various groups we know today their lice diverged with them. The closest relative of the human body and head lice is the chimpanzee louse, *Pediculus schaeffi.* The human body and head lice and the chimpanzee louse diverged from one another about six million years ago when the human and chimpanzee evolutionary lines diverged.

The evolutionary relationship between the head louse and body louse is very interesting because it appears the human body louse diverged from the

Head lice at various stages of development from egg to adult. (iStockPhoto)

human head louse sometime between 30,000 and 114,000 years ago—a blink of an eye in evolutionary terms. As our ancestors evolved, their body hair became finer and finer, which afforded less and less protection to unwanted passengers, such as lice. The lice sought refuge in the thick hair of the scalp, which remains to this day, but sometime between 30,000 and 114,000 years ago, it appears our ancestors—probably in the cooler, northern latitudes—began wearing clothes on a regular basis. Garments presented the lice with a new place to live and the ancestors of the body lice (the head lice) took advantage of this new habitat.

The long, albeit unwanted, relationship that lice have with humans has embedded them in our cultural heritage to the extent where they have found their way into everyday language. Common words and phrases, such as *lousy, nitwit, nitty-gritty, nitpicking,* and *going over with a fine-tooth comb* all relate to the lice species that live on us. Before the advent of synthetic insecticides and their widespread usage, lice were considered to be an unpleasant, occasionally deadly fixture of human life; parasites with no respect for class or wealth. Everyone had lice and it was simply a case of tolerating them and getting on with things. Today, when children come home from school with head lice it is seen as something of a social stigma, even though there's no correlation between small infestations of these parasites and hygiene.

Female body lice and head lice can produce as many as 300 eggs (nits) in their lifetime, each of which is attached to a hair or fiber of clothing with an adhesive secretion. When the nymphs hatch they are able to suck blood almost immediately, developing rapidly and reaching maturity in as little as nine days. Body lice stay in the clothing, only venturing onto the body to feed, but head lice always stay on the scalp and are commonly found on the back of the neck and behind the ears. To feed, these insects insert their tube-like mouthparts through human skin into a capillary where blood can be extracted. In normal circumstances the lice drink 12.5–25 percent of their body weight in blood every 4–6 hours and in the event they are separated from their host or cannot get to the skin they die after 20–48 hours.

Lice are considered to be pests because their bloodsucking and the irritation it causes can be annoying. More insidiously, they also transmit pathogens. In heavy infestations of body lice the feeding activities of hundreds or thousands of these insects over long periods of time can cause a condition known as vagabond's disease, characterized by darkening and thickening of the skin. Where head lice are left untreated and in the absence of hair care (washing and brushing), a condition known as plica polonica, or Polish plait, can develop where the hair becomes matted with skin exudates, blood, and the feces and secretions of lice, eventually forming a thick mass of foul-smelling hair harboring large numbers of lice and their assorted waste. In days gone by, vagabond's disease and plica polonica were commonplace and not restricted to the peasantry. Common medieval diseases aside, the bites of lice can be intensely itchy and scratching them causes inflammation of the skin and secondary infections if the skin is broken.

In addition to the irritation caused by their bites, lice can transmit several pathogens, some of which cause serious disease. Of these, the bacterium *Rickettsia prowazekii* is the most important as it causes epidemic typhus (also known as louse-borne typhus). Only body lice transmit this bacterium and unusually for an insect-borne pathogen the infection is fatal to the louse as the microorganisms invade and rupture the cells of the gut. Humans are infected with this disease when pathogens find their way onto the skin either via lice feces or when these insects are inadvertently crushed against the skin. From here the bacteria invade the tissues through a wound, such as skin breaks made by scratching lice bites. Once in the human body the bacteria cause a rash, fever, profuse sweating, and without treatment, nervous system symptoms and death. In the past, epidemic typhus was a real menace and it has certainly been responsible for or

Lice Species Feeding on Humans and Domesticated Animals

Species	Host
Chewing lice (Mallophaga)	
Shaft louse (*Menopon gallinae*)	Domestic fowl
Yellow body louse (*Menacanthus stramineus*)	Domestic fowl
Fluff louse (*Goniocotes gallinae*)	Domestic fowl
Brown chicken louse (*Goniodes dissimilis*)	Domestic fowl
Wing louse (*Lipeurus caponis*)	Domestic fowl
Chicken head louse (*Cuclotogaster heterographus*)	Domestic fowl
Large turkey louse (*Chelopistes meleagridis*)	Turkeys
Slender turkey louse (*Oxylipeurus polytrapzius*)	Turkeys
Slender pigeon louse (*Columbicola columbae*)	Pigeons
Duck lice (*Anaticola crassicornis* and *A. anseris*)	Ducks
Gyropus ovalis	Guinea pigs
Gliricola porcelli	Guinea pigs
Heterodoxus spiniger	Dogs
Bovicola bovis	Cattle
B. equis	Horses, mules, and donkeys
B. ovis	Sheep
B. caprae	Goats
Trichodectes canis	Dogs
Felicola subrostratus	Cats
Sucking lice (Anoplura)	
Head louse (*Pediculus humanus capitis*)	Humans
Body louse (*Pediculus humanus humanus*)	Humans
Crab louse (*Pthirus pubis*)	Humans
Hog louse (*Haematopinus suis*)	Pigs
Short-nosed cattle louse (*H. eurysternus*)	Cattle
Cattle tail louse (*H. quadripertusus*)	Cattle
H. tuberculatus	Water buffalo
H. asini	Horses, mules, and donkeys
Linognathus spp.	Cattle, sheep, goats, and dogs
Solenopotes capillatus	Cattle
Polyplax spinulosa	Rats

contributed to many epidemics through the ages. In Mexico during 1576, 2 million deaths (of a population of 9 million) were attributed to epidemic typhus. Between 1917 and 1921 there were at least 25 million cases of typhus in the Soviet territories, resulting in 0.5–3 million deaths. These examples demonstrate what a serious disease louse-borne typhus once was. Fortunately, today, insecticides, antibiotics, and vaccines have limited the potency of this disease, but it has not been eradicated and in the event of a large-scale war or natural disaster it would return with a vengeance.

The other two important diseases transmitted by human lice are trench fever and relapsing fever. Trench fever is caused by the bacterium *Rochalimaea quintana,* and although it's nonfatal it can still be very debilitating, causing a rash and long-lasting fever. Like epidemic typhus, the bacteria are spread in louse feces or when a louse is crushed and it gets into the body via wounds or inhalation. Relapsing fever is caused by a spirochete bacterium (*Borrelia recurrentis*) and it too infects humans when a louse is crushed and the pathogens gain entry through a wound; however, these bacteria may also be able to penetrate unbroken skin. Epidemic typhus, trench fever, and relapsing fever often occur together, especially in poverty-stricken populations with poor hygiene and where events such as wars, famine, and natural disasters produce conditions that are conducive to the spread of the bacteria that cause these diseases.

In view of the human health impacts of lice-borne diseases, a long war has been waged against these insects. In relatively recent decades, the widespread use of insecticides and an increased understanding of louse and louse-borne pathogen biology have lessened the impact these insects have on humankind. Current strategies for controlling these insects relies on insecticide sprays and dusts, monitoring, and making sure that garments infected with the body louse are thoroughly cleaned. It's worth remembering that our control of these insects is tenuous at best. Wars, natural disasters, and insecticide resistance could easily conspire to render these insects a serious threat to human health once more.

FURTHER READING

Durden, L. A., and J. A. Lloyd. Lice (*Pthiraptera*). In *Medical and Veterinary Entomology* (G. R. Mullen and L. A. Durden, eds.), pp. 56–80. Academic Press, San Diego, CA, 2009.

Kittler, R., M. Kayser, and M. Stoneking. Molecular evolution of *Pediculus humanus* and the origin of clothing. *Current Biology* 13(16)(2003): 1414–17.

Horse Botflies

Horses are incredibly important animals, even today. It is estimated there are around 58 million horses in the world. Although nowadays in the developed world they are kept for recreation rather than out of necessity, there are still plenty of places around the globe where these animals are an integral part of everyday life as both beasts of burden and a source of animal protein. The economic importance of horses is hard to quantify, but it has been estimated that in the United States alone (a country with 9.5 million of these animals), the horse industry is directly worth $39 billion every year and the overall value of horses is estimated to be around $102 billion.

These are colossal sums, so it's perfectly logical that any organism that impacts the well-being of these infinitely useful animals will be deemed a pest. One such group of animals are the horse botflies, insects with some rather grisly habits that have accompanied horses in their spread around the world. Three important species of horse botfly are known—*Gastrophilus intestinalis* (horse botfly), *G. nasalis* (throat botfly), and *G. haemorrhoidalis* (nose botfly). In outward appearance these flies resemble honeybees, but this is no more than a superficial resemblance as the horse botflies are interested in equids rather than flowers and the production of honey.

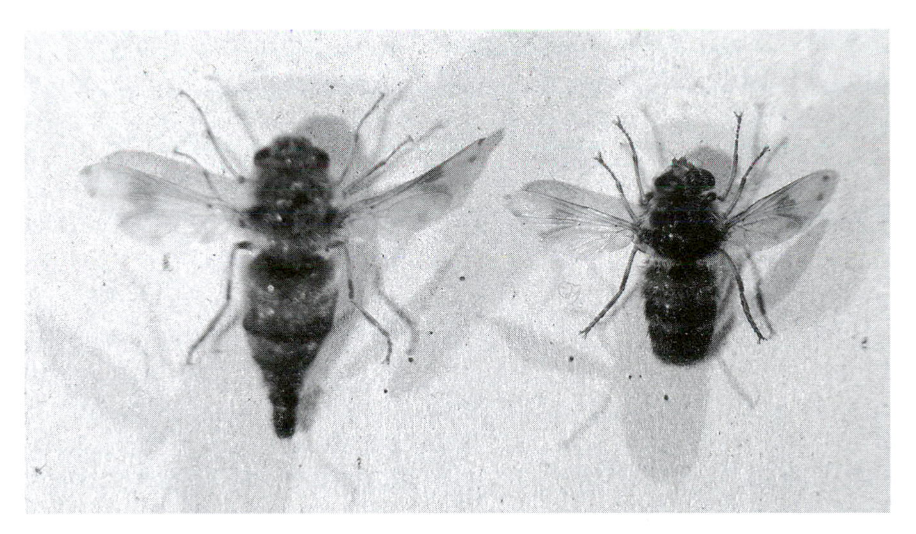

A pair of adult horse botflies. (iStockPhoto)

Females of these flies harass horses, using their excellent flying abilities to hover over their quarry before embarking on dive-bombing sorties. Almost as if the horses understand the significance of these particular buzzing insects and what it means if they don't get away, they run and buck in an effort to evade or repel the flies. Unfortunately for the horses, the flies are persistent, and sooner or later a female manages to alight on her host to deposit her eggs. The females lay between 150 and 1,000 eggs, sticking each one to the base of a hair to prevent it being accidentally dislodged by the horse. The three species mentioned above deposit their eggs in different places. *G. intestinalis* goes for the hair around the knees, while *G. nasalis* and *G. haemorrhoidalis* go for the hair under the jaw and the lips, respectively. In the first species, egg hatching is triggered by the warmth and saliva of the horse licking the hair around its knees. Upon hatching, the larvae penetrate the epithelium of the tongue and tunnel their way down into the animal's stomach, where they emerge and latch onto the stomach mucosa with their strong mouth hooks. Just how these larvae are able to tolerate the intensely acidic environment of the horse's stomach is something of a mystery, but here they stay, rasping at the stomach lining and ingesting the blood that flows out of the small wounds. They remain securely attached to the stomach lining for around 10 months until the following spring/early summer. At this time they relinquish their grip and get carried through the animal's digestive tract, eventually emerging in the horse's feces. Now ready to pupate, the larvae burrow into the soil and begin the transition into adulthood.

The throat and nose botflies also take up residence in the stomachs of their hosts, but the eggs of these species just hatch in their own time (four to five days), whereupon the larvae enter the mouth between the lips and tunnel into the epidermis to continue their journey to the stomach. The nose botfly differs from the other two species in that the third instar larvae attach themselves to the horse's anus for a short time before passing out and completing their development. The significance of this anomalous behavior is not understood.

Throughout their range the prevalence of botfly infection among horses can be high. For example, an investigation of 725 horses in Poland found that 104 animals (14.8%) were infected with *Gastrophilus* larvae. The numbers of larvae found in each infected horse ranged between 1 and 925 with a mean of 52 per horse. In other countries the prevalence of horse botfly infection can be anywhere between 9 and 99 percent. The economic losses attributable to these interesting parasites have not been quantified. It seems that in small numbers, stomach botflies do not cause horses any significant problems, but although there may be no outward signs of infection it is possible the tunneling and feeding activities of the larvae permit the entry of pathogens that may cause disease later in the animal's life. Heavy infections of horse botfly are quite a different matter as they can cause significant damage to the lining of the stomach and the intestine with their feeding activities. Large numbers of larvae can also result in the blockage of the pyloric valve, the narrow conduit between the stomach and small intestine. Such a blockage can be fatal. Large numbers of these larvae feeding in the stomach for a period of several months will ingest a considerable amount of blood, which can weaken the horse and make it more vulnerable to disease. There is also the issue of the damage the first instar larvae can do to the mouth, tongue, and throat as they penetrate the epidermis and burrow toward the stomach.

Another often overlooked impact of these flies is the stress they can cause horses during the female's oviposition activities. The horses, disturbed by the presence of these insects, will try and escape, which results in them getting injured and spending less time grazing.

The horse botflies are native to the Old World, but they have been transported with horses all over the world. Control of these flies is difficult because outward signs, other than eggs on the legs and head, are uncommon and easily overlooked. A simple way of limiting the populations of these flies is removing or killing the eggs as soon as they are spotted. This can be done simply and quickly with grooming, warm water, or various chemical solutions. Pesticides (e.g., ivermectin and moxidectin) primarily intended for intestinal worms are also known to be effective against the horse botflies and the widespread use of these chemicals during recent decades is one reason why these flies are not as common as they once were. Another simple means of controlling these flies is ensuring that horse feces in enclosed areas such as paddocks are quickly removed and discarded before the larvae have a chance to escape into the soil to pupate.

FURTHER READING

Catts, E. P., and G. R. Mullen. Myiasis *(Muscoidea* and *Oestroidea).* In *Medical and Veterinary Entomology* (G. R. Mullen and L. A. Durden, eds.), pp. 318–49. Academic Press, San Diego, CA, 2009.

Housefly

The housefly is one of the most ubiquitous insects and also one of the most loathed. An animal of decaying matter, the housefly can exploit any suitable food resources quickly and effectively. The larvae (maggots) will develop rapidly in just about any organic waste, from the accumulated feces in farmyards to the decomposing organic matter at the bottom of a garbage can. Animals like the housefly are instrumental in the successful functioning of terrestrial ecosystems because they hasten the breakdown of waste and dead organisms, returning some of the building blocks of life back to the soil or channeling them back into the food web. The flies, as a group, are nature's garbage men and if it were not for them the earth would be buried beneath a sea of decaying matter.

The problem with the housefly is that it has taken to living in association with humans simply because we are wasteful: wherever there are humans there's a surfeit of food for any self-respecting, filth-loving insect. The long association of the housefly with humans probably began when family

Disease-causing organisms find their way from rotting matter to human food on the many hairs and bristles and in the gut of the housefly. (Risto Hunt | Dreamstime.com)

groups of our ancient ancestors sought refuge from the elements, initially in natural shelters, such as caves and then in purposely built dwellings. The propensity of the housefly to explore dark and dim places suggests this species was originally associated with caves, as they may have been exploiting the constant temperature and humidity and decaying matter in these natural refuges. A female housefly, heavy with eggs, is drawn to the heady odor of decaying organic matter and she deposits her eggs directly on the substrate that will nourish her offspring. Each female housefly produces 120 to 150 eggs in at least six batches. From these eggs pallid maggots hatch to commence the race to grow, pupate, and mate, thus completing the cycle. The maggots are unparalleled eating machines and because their sole purpose is to grow and accumulate energy to enable the shift through pupation into adulthood, they are structurally rather simple. Senses are an extravagance when you don't need to find food, so these are rudimentary and limbs of any kind would be a hindrance tunneling around in filth, so they don't have any. What they do have is a powerful tubular body to aid them in their explorations of their food and a tough hook to bring edible matter into their mouth. In suitable conditions it takes as little as 10 days for the housefly to develop from egg to adult. With such reproductive and growth potential it is easy to see how huge populations of housefly can quickly develop.

Houseflies have a penchant for decaying matter, so the adults come into contact with viruses, bacteria, fungal spores, protozoa, and parasite eggs and end up carrying them around in two ways. Firstly, the adult flies feed on decaying matter to fuel their flight and to provide the nutrients that will complete the maturation of their sex cells. They lack chewing mouthparts, so they are dependent on liquid food, which they obtain with the unsavory technique of vomiting digestive juices onto their food via their proboscis. The resultant soup is sucked back up complete with whatever pathogens were present on the decaying matter the fly was feeding on. The next time the fly feeds, some of these pathogens will be regurgitated. Yet more pathogens will find their way into the insect's feces, possibly contaminating uncovered food destined for humans and animals. Not only does the adult fly inadvertently swallow pathogens, but its body is covered with a multitude of tiny scales and bristles that collectively act as a sponge for viruses, bacteria, fungal spores, protozoa, and parasite eggs. In particular, the tiny feet that afford these insects such an excellent grip on walls and ceilings are veritable pathogen magnets that spread all sorts of potentially disease-causing organisms over whatever surface the flies scuttle across. The pathogens and parasites on the fly's surface and those in its regurgitate and feces all contribute to the trail of contamination left by these insects.

Houseflies are known to carry at least 100 different pathogens and they are vectors for at least 65 of these. Some of the more important pathogens transmitted by these insects include the viruses that cause polio and hepatitis; the nematodes commonly known as thread worms (*Trichuris* sp.) and hook worms (*Ancylostoma* sp.); the protozoa (*Entamoeba* sp.) responsible for amoebic dysentery; and the bacteria that cause salmonellosis, diphtheria (*Corynebacterium diphtheria*), and tuberculosis (*Mycobacterium* sp.). Most of the pathogens transmitted by houseflies are picked up when the adult flies feed on feces and then contaminate fresh food or water.

Houseflies are of considerable importance to human and animal health. As the human population continues to expand and habitable areas become increasingly overcrowded, the impact of this insect on humans and livestock will become more intense. Currently, water and waste infrastructures are just about managing to cope with the burgeoning human population, but in the event of natural disasters and wars, housefly-borne diseases can be devastating as food and water supplies are contaminated.

Controlling houseflies and their potentially devastating impact on human and animal health can be relatively straightforward, but their sheer numbers and resilience can make it a very expensive and time-consuming endeavor. The most effective way of limiting the population size of house flies is denying them suitable breeding sites. Therefore, exposed refuse should be buried or held in containers that prevent the entry of female houseflies laden with eggs. Adult flies can be caught and killed with various traps and insecticides can also be used when there are enormous populations to try and control. New approaches for controlling these insects include insect-killing fungi that can be distributed in their spore stage to infect and kill adult house flies, and various parasitic wasps that use housefly maggots as food for their own larvae. Compared to insecticides, these methods are environmentally friendly, but they cannot eradicate a fly population. In situations where it is not possible to effectively control the populations of houseflies, various preventative measures can be used to ensure the flies are not able to contaminate food with the varied pathogens they are capable of transmitting. One simple example is making sure that food is always covered when adult houseflies are around.

FURTHER READING

Malik, A., N. Singh, and S. Satya. House fly (*Musca domestica*): A review of control strategies for a challenging pest. *Journal of Environmental Science and Health Part B* 42(2007): 453–69.

Japanese Beetle

This insect, *Popillia japonica,* is a handsome species belonging to the group of scarab beetles known as chafers. The native range of this beetle is Japan, northern China, and the far east of Russia, but in relatively recent times it has found its way to Portugal, Canada, and the United States. In all the areas where it has been introduced, it has proceeded to make a nuisance of itself by damaging ornamental plants, turf, and crop plants on a large scale. It was first detected in the United States in 1916 in New Jersey. Since then it has spread rapidly through many of the U.S. states east of the Mississippi River, with the exception of Florida, where the climate is a little too subtropical for this decidedly temperate insect.

Like all beetles, the appearance of the rather handsome adult is in stark contrast to the unfortunate larva—a corpulent pale grub with a face fit for radio. Almost as if it's ashamed to show itself in mixed company, the larva lives out its entire life underground, nibbling at the roots of grasses for many months until it's ready to have its insides and outsides reordered in the rigors of metamorphosis. When the adult hatches from the pupa, clothed in lustrous chitin, it tunnels its way to the surface and takes to the wing, leaving the dark, dank confines of its subterranean way of life behind forever. Chafers are strong albeit clumsy fliers, and they commonly fly into vegetation, buildings, people, and anything else in their flight path. The adult lives rather a fleeting existence and in most cases it has just over a month to find a mate and copulate, thus safeguarding the passage of its genes into the next generation. During its brief adult existence the Japanese beetle is a very unfussy eater and will quite happily eat the leaves and flowers of at least 300 species of plant. Crops grown for human consumption commonly damaged by this beetle include asparagus, soybean, apple, *Prunus* species, rhubarb, roses, *Rubus* species, grapes, and corn, to name but a few. Where dense aggregations of these beetle occur, it's not unusual to see fruit trees and other plants stripped of their foliage and fruits. Many types of ornamental plant grown by horticulturalists for gardening and landscaping are also ravaged by these beetles. There is also the

The Japanese beetle can be a serious pest of ornamental plants and gardens, especially in areas where it has been accidentally introduced. (Bruce Macqueen Dreamstime.com)

feeding activity of the larvae, which does nothing to endear this species to agriculturalists and horticulturalists. The grubs feed on the roots of grasses and in doing so they can cause considerable damage to pastures and lawns. The quality and the appearance of the sward on golf courses and the perfectly manicured lawns of parks and those surrounding important buildings is of paramount importance to many organizations, so any animal that damages this green carpet is treated with disdain. The root damage inflicted by the chomping larvae of Japanese beetles can be enough to prevent grasses from absorbing sufficient moisture and nutrients, resulting in large, brown, withered patches where the grass has died.

The ability of this beetle to damage a wide variety of crops makes it a pest of considerable economic importance in the areas in which it is now found. Its populations outside its natural range can become very large in the absence of natural enemies in areas where it has been introduced. It has been estimated that management of this pest costs the turf grass and ornamental plant industry at least $450 million every year in the United States alone. This estimate does not take into account the damage caused to crops, and the dollar loss in yields and various strategies that have been implemented to control its westward spread.

Controlling this beetle is quite a challenge because the larvae spend almost their entire life underground, where they are protected to a certain degree from the toxic effects of insecticides. Furthermore, the adults are powerful flyers able to disperse over significant distances to find new areas of habitat. Traditionally, liberal application of insecticides has been the favored approach, but as our understanding of the full extent of the effects of these chemicals has increased, scientists and growers alike are searching for more environmentally friendly means of controlling these beetles. Insecticides, especially the nonpersistent, more specific compounds, still have a role to play, but as complements to other strategies.

This beetle is a large insect, so one simple and cheap method of control is searching plants and removing the adults for later disposal. Nets and other barriers can also be used to exclude the beetles from plants susceptible to damage. Both of these low-tech approaches are very effective on the small scale, but in large infestations they are just not practical. Quarantine on the movement of plants can be effective at blocking the spread of this beetle, especially as immature stages hiding away in the soil can be detected. Aggregation pheromones are important to these beetles, especially when they are newly emerged and are seeking out others of their kind. Using these pheromones to bait traps into which the beetles are lured has been investigated, but it seems that it is more effective as a means of assessing the level of an infestation rather than a way of controlling their numbers.

There are many organisms that parasitize and infect the various life stages of the Japanese beetle. Many of these are being investigated as biological control agents. There are the wasps, *Tiphia vernalis* and *T. popilliavora,* that parasitize the subterranean larvae of this beetle. The larvae are also known to be infected and killed by several parasitic nematodes. These nematodes can be applied to infested areas and are as effective as insecticides at controlling an infestation. There is also a parasitoid fly, *Istocheta aldrichi,* that attacks and kills the adult beetles.

The eggs, larvae, pupae, and adults of this beetle are also predated on by a number of other animals, including a variety of arthropods and many vertebrates. Enhancing cultivated environments to make them conducive to the survival of all these natural enemies provides a safe, sustainable means of regulating the populations of the Japanese beetle, so that potentially destructive outbreaks do not occur.

FURTHER READING

Potter, D. A., and D. W. Held. Biology and management of the Japanese beetle. *Annual Review of Entomology* 47(2002): 175–205.

Kissing Bugs

Kissing bug conjures up images of a gentle creature with behavior reminiscent of our romantic gestures, but how wrong such an assumption would be. There's nothing romantic or particularly charming about the large bugs belonging to the family *Reduviidae.* In fact these insects and their way of life are sure to give most people the jitters. The three most important species of kissing bug are *Triatoma infestans, T. dimidiata,* both of which can be more than 3 centimeters long, and the much smaller *Rhodnius prolixus.* Like their relatives the bedbugs, kissing bugs are blood feeders and during the day they hide away in suitable cracks and crevices, often in and around human dwellings. At night they emerge to suck blood. The name *kissing bug* relates to the fact that they often suck blood from the around the lips of sleeping humans. Like other insects that take blood from large animals, the kissing bugs do so painlessly; therefore the slumbering victim is often unaware of having one or more of these considerable insects attached to their face. Feeding takes anywhere between 3 and 30 minutes and like the other blood-sucking bugs they depend on symbiotic bacteria contained within the lining of their gut for some of their nutrition.

In rare cases, severe reactions to kissing bug bites can occur, but the main concern with these bugs is that they transmit Chagas disease, a devastating disease caused by the protozoan *Trypanosoma cruzi.* This is a disease of the Americas, primarily South and Central America, although the disease is a growing concern in the United States as immigration from areas where it is endemic increases. Figures from 2006 show that at least 8 million people are infected with Chagas disease; in 2008 the global death toll was at least 11,000. The protozoan that causes Chagas disease is present in the feces of the kissing bugs and as they feed they defecate on the

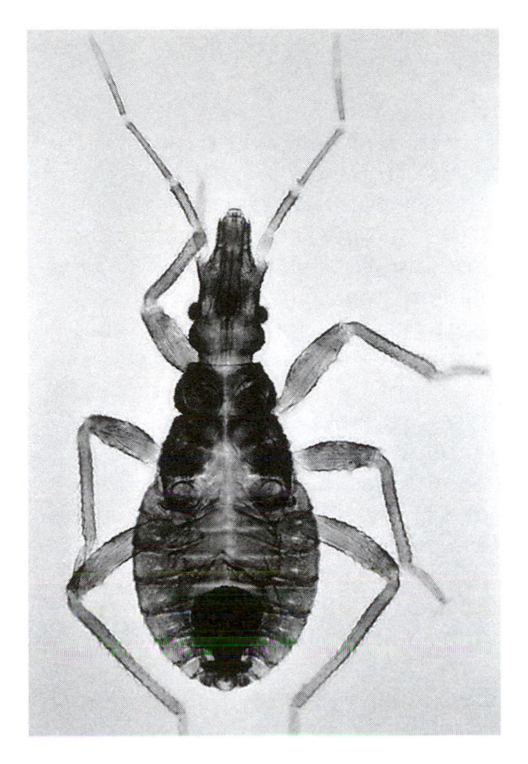

A young *Rhodnius prolixus* nymph, one of the kissing bug species that is a vector of Chagas disease. (CDC/Donated By the World Health Organization, Geneva, Switzerland)

victim's skin. The protozoan gets into the body through the insect's bite, a small wound, or the permeable tissue of the mucous membranes (e.g., the eyes and mouth) when these are rubbed by fingers contaminated with the insect's feces. Like the organism responsible for causing malaria, the life cycle of the Chagas disease parasite is very complex, but they end up invading all the tissues in the body, with a preference for nerve and muscle cells. The symptoms are varied, including anemia, weakness, nervous system disorders, chills, muscle and bone pain, enlarged colon, esophagus, and heart. It is infection of the heart tissue that accounts for the majority of deaths caused by Chagas disease. The parasite destroys the nerves of the heart, so the organ loses muscle tone, becomes weak, and increases in size to such an extent that it can no longer pump blood effectively. Drugs are available to treat Chagas disease, but they are not completely effective and the more widely they are used the faster resistance will evolve in the protozoan parasite.

Although Chagas disease can be transmitted between humans in a number of ways, such as blood transfusions, the kissing bugs are the most

The Evolution of Bloodsucking Insects

Lots of insects suck the blood of vertebrates, but how did this interesting way of life evolve? Blood is rich in protein, which is crucial for egg maturation in many insects. Also, there's a lot of blood about as long as you can get at it, so there was selection pressure for the evolution of structures and behaviors that made this possible. To understand the evolution of bloodsucking we have to think about the ways in which smaller animals, namely insects, associate themselves with larger animals. Firstly, an insect may live alongside a vertebrate because it gets access to food or shelter—we can view this stage as commensalism and it is the first rung of the ladder that reaches to parasitism. Over time and countless generations these insects may start to live on the vertebrate, perhaps nibbling shed skin, scales, or feathers. These insects are now parasites as they're living directly on the host. More stretches of time pass and gradually the insect assumes an even more parasitic way of life: feeding from the around the edges of wounds or opening the skin directly to get at the fluids beneath. The path is now clear for the evolution of forms that can penetrate the skin directly to suck blood without irritating the host too much and attracting unwanted attention.

This is one way in which bloodsucking may have evolved in insects. The second way is similar, but involves the twist of preadaptation, that is, the ancestor of the bloodsucking insect has characteristics or behaviors that make the switch to sucking vertebrate blood a relatively small one. Kissing bugs are examples of preadapted creatures as their ancestors were very probably insect predators, much like the modern-day assassin bugs, which brandish robust, elongated mouthparts to pierce the tough exoskeletons of their invertebrate prey. The direct ancestors of kissing bugs were probably assassin bug-like creatures that spent increasing amounts of time in and around the nests of birds and mammals because of the various scavenging and parasitic insects that are to be found there. A few speculative jabs at the nest builder with their pointy mouthparts put these bugs on the road to becoming obligate vertebrate bloodsuckers, and we have the situation where a species that started out as beneficial (preying on nest scavengers and parasites) evolves into a harmful parasite.

As soon as the leap to feeding on vertebrate blood has been made, any vertebrates become potential sources of food, which is why bloodsucking insects often feed on a large range of hosts. Interestingly, one step in the evolution of kissing bugs from nest predators to human blood feeders can be still be seen today in the mountains of Bolivia. There, populations of *Triatoma infestans* feed exclusively on wild guinea pigs, which make their nests in rock piles. If humans were to build dwellings nearby it would not take long for the bugs to move in and begin their nocturnal bloodsucking.

Further Reading: Schofield, C. J. Biosystematics and evolution of the Triatominae. *Cadernos Saúde Pública* 16(Sup. 2)(2000): 89–92; Lehane, M. J. *The Biology of Blood-Sucking in Insects.* Cambridge University Press, Cambridge, MA, 2005.

important link in the transmission of this condition. Therefore controlling these insects ultimately holds the key to controlling and possibly eradicating this potentially lethal disease.

Simple measures to reduce the populations of kissing bugs are surprisingly effective and can include improving the finish of dwellings by using plaster and metal roofs to deprive the insects of their daytime refugia. These simple measures combined with fumigation and the treatment of infected humans can reduce the impact of this disease considerably. For example, the global death toll from Chagas disease in 1990 was 45,000, compared with 11,000 in 2008; and the estimated number of infections in 1990 was 30 million, which had fallen to 8 million by 2006. As significant as this progress is, Chagas disease is far from being on the ropes. Global travel and immigration means the disease is now found beyond its traditional borders, including parts of the United States and Europe, and emergent cases in areas previously considered to be free from the disease are cause for concern. Drug and insecticide resistance in the protozoan and the kissing bugs, respectively, has also been observed and this is something that will only get worse in the future.

FURTHER READING

Lehane, M. J. *The Biology of Blood-Sucking in Insects.* Cambridge University Press, Cambridge, MA, 2005.

Locusts

"Thou shalt carry much seed out into the field, and shalt gather but little in; for the locust shall consume it." As this quote from the Bible shows, locusts have been a problem for humankind for a very long time indeed. It is very likely that locusts have been nibbling crops destined for humans and livestock ever since our species began cultivating plants in the Fertile Crescent at least 10,000 years ago.

Locusts are technically grasshoppers that go through intermittent population explosions, forming swarms. Several species of grasshopper are commonly known as locusts and all of them can be considered crop pests to greater or lesser extents (see sidebar).

Locusts begin life as nymphs—small, wingless miniature adults. To grow, the nymphs shed their skin a number of times. The last time they go through this process they emerge with shriveled wings that need to be pumped full of blood and left to harden for a while before they are ready for flight. Adult locusts divide their time between eating and trying to mate. As soon as the female locust's eggs are fertilized, she deposits them in sandy soil, often cocooned in a foam that quickly hardens to protect the eggs while the young are developing. Certain characteristics of locust biology make them perfect pests. Firstly, they are generalist herbivores, able to consume just about any part of the plants they eat, and secondly they are prolific breeders capable of building large populations very quickly.

Probably the most serious locust pest is the desert locust, a species that often forms huge swarms capable of devastating crops in some of the poorest places on earth. Like all locusts, the desert locust has a solitary phase and gregarious phase, and it is the latter that is the problem. In normal situations the desert locust will be present in the solitary phase, but following periods of drought, deluges may promote the growth of abundant, albeit transient, vegetation. To exploit this greenery the population of the desert locust explodes and in a short space of time there can be dense aggregations of these insects busily munching all the available plant matter. Making use of the abundant food on offer and the high density of the

The Locust Species and Their Geographic Distribution

Locust species	Geographic location
Migratory locust (*Locusta migratoria*)	Africa, Asia, and Australasia
Red locust (*Nomadacris septemfasciata*)	Tropical Africa
Desert locust (*Schistocerca gregaria*)	Africa, Middle East to India
Brown locust (*Locustana pardalina*)	South Africa
Plague locust (*Chortoicetes terminifera*)	Australia
Bombay locust (*Patanga succincta*)	India to China
Mediterranean locust (*Dociostauras maroccanus*)	Mediterranean
American locust (*Melanoplus* spp.)	United States and Canada

insects stimulates changes resulting in the development of gregarious locusts that look and behave differently from their solitary kin. Before too long, the vegetation begins to wither and die and the gregarious locusts are faced with the prospect of starvation, so they move en masse, either on foot or on the wing, to search for more food. Swarms dotted about the landscape band together to form huge assemblages of locusts that scour the land looking for food. The size of these swarms is incredible, as there can be billions of individuals at a density of 80 million per square kilometer covering an area of 1,000 square kilometers (these swarms are massive, but they pale into insignificance alongside the aggregations of the now extinct Rocky Mountain locust observed in the United States and Canada during the 19th century—see sidebar). Desert locust swarms can cover significant distances in a day, with winged adults being able to travel around 100 kilometers, while the flightless nymphs are able to march around 1.5 kilometers each day.

Why locusts should form these enormous aggregations is not entirely understood, but perhaps there are advantages in finding food as many millions of eyes will be more likely to find new areas of lush vegetation than just a few. Whatever the advantages are of swarm formation, these insect collectives are adept at finding and exploiting new food resources, and very few cultivated plants are free from the munching jaws of the locust. These insects are content to eat the leaves, stems, flowers, bark, fruit, and seeds of valuable plants, such as millet, rice, banana, wheat, barley, date palm, vegetables, and maize. Not only are they unfussy when it comes to food, but locusts also have very big diets. An adult desert locust weighs about 2 grams and it can eat its own body weight in food every day; therefore a swarm of one billion locusts can consume 2,000 tonnes of plant matter every day, much of which was intended for humans and livestock.

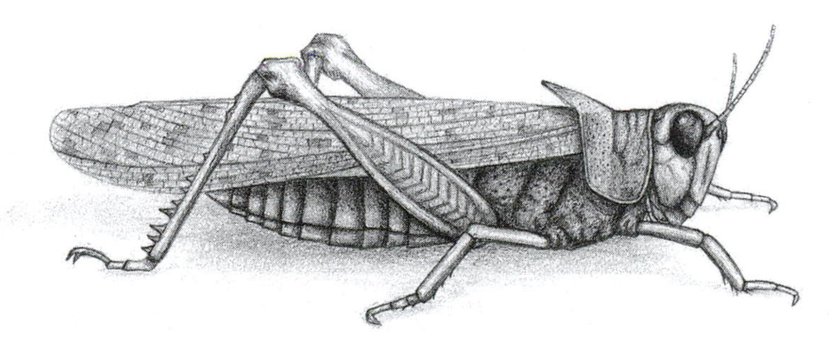

Huge numbers of the Rocky Mountain locust once swarmed in North America, but they became extinct by the beginning of the 20th century. (Phil Miller)

Although a huge amount has been written about locusts, figures that quantify how much damage they actually do is remarkably scarce. Some reports include the $50 million worth of crops lost to a locust outbreak in Morocco in 1954 and the loss of around 150,000 tonnes of grain in Ethiopia in 1958—sufficient to feed approximately one million people for one year. More recently, the desert locust is estimated to have caused crop losses to the tune of $2.5 billion between 2003 and 2005 in West Africa alone.

The scale of the locust problem means that considerable amounts of time and money have been ploughed into attempts to control them; unfortunately, these efforts have yet to dent the locust swarms when they

The Mysterious Disappearing Locust

The Rocky Mountain locust was once found in immense aggregations. A swarm observed and monitored in Nebraska during 1874 was estimated to be 2,900 kilometers long and 1,800 kilometers wide, a fluttering mass of some 12 trillion insects with a combined weight of 27 million tonnes that was said to take five days to pass overhead. If the desert locust is anything to go by, this enormous swarm of Rocky Mountain locusts may have required its own weight in food every day. A mere 30 years after this enormous swarm and others devastated crops in the Midwest of the United States, this insect had disappeared. Exactly how such a numerous insect became extinct has been a bone of contention for some time, but the likely explanation is that outside of its swarming periods the locust retreated to the sheltered valleys of Wyoming and Montana, where the females laid their eggs in the fertile soil. These very same valleys attracted the attention of settlers who saw their potential for agricultural pursuits and with their horses and their ploughs they turned the soils over and grazed their livestock on the nutritious grass. These actions destroyed the eggs and developing young of the insect and around three decades after its swarms blotted out the sun, the Rocky Mountain locust was gone forever.

Further Reading: Lockwood, J. A., and L. D. DeBrey. A solution for the sudden and unexplained extinction of the Rocky Mountain grasshopper (Orthoptera: Acrididae). *Environmental Entomology* 19(1990): 1194–1205; Lockwood, J. A. Voices from the past: What we can learn from the Rocky Mountain locust. *American Entomologist* 47(2001): 208–15; Lockwood, J. A. *Locust: The Devastating Rise and Mysterious Disappearance of the Insect that Shaped the American Frontier.* Basic Books, New York, 2004.

appear. A great deal of research has been directed at gaining an understanding of the factors that contribute to locust swarming so these population explosions can be predicted and nipped in the bud. The Desert Locust Control Organization (DLCO) was set up in 1960 to oversee the monitoring and control of the desert locust, coordinating the efforts of a number of countries to monitor the desert locust populations and limit the spread of outbreaks when they can't be contained. The effectiveness of the DLCO is often hampered by political problems among the participating countries, but it nonetheless improves the desert locust problem. When locust outbreaks do occur a number of methods are brought to bear to kill the insects. Low-tech methods include the construction of trenches to trap marauding nymphs, and creation of smoke, which is said to repel the swarms. Insecticides are the standard response in large outbreaks and huge quantities are sprayed from the air to kill locusts. For example, the last major outbreak of desert locusts in Africa in 1988 saw the spraying of 1.5 million liters of insecticides—an enormous quantity, the long-term environmental consequences of which are unimaginable. Not only are such strategies environmentally unsound, but they are extremely expensive: this 1988 control effort cost $300 million. The expense and environmental toxicity of conventional insecticides inspires a great deal of interest in other ways of controlling locust outbreaks, including formulations containing fungal spores that infect and kill locusts, extracts from various plants, and insect growth regulators.

FURTHER READING

Capinera, J. L. *Encyclopedia of Entomology*, Vol. 4. Springer, Dordrecht, Germany, 2008.

Hill, D. S. *The Economic Importance of Insects.* Chapman & Hall, London, 1997.

Mealybugs

The mealybugs are very closely related to the scale insects, which have been covered in a separate entry. Taxonomically, the mealybugs are in their own family, the pseudococcidae, and it is very likely these animals are ancestral to the scale insects as they possess many of the traits exhibited by these bizarre, largely sedentary animals, albeit in a more primitive state. The major difference between the mealybugs and the scale insects is their

A mealybug on a cassava stem from an affected plantation in northeastern Thailand. (AP/Wide World Photos)

degree of mobility. The mealybugs are much more mobile than scale insects and they retain fully functioning legs and antennae to move around their host plant and locate the best feeding sites.

Like scale insects, the mealybugs also produce a waxy substance, which is likened to the meal obtained from grinding seeds and grain, hence their common name. This fluffy wax adheres to the body of the mealybug, lending it the appearance of an animated crumb. The life cycle of mealybugs is very similar to the scale insects. The first instar, mobile nymphs, hatch from tiny orange eggs and these crawlers disperse around the host plant to locate suitable feeding sites. When they become more settled they start to secrete the waxy material that will cover their entire body. The nymphs shed their skin twice and in doing so they steadily take on the adult form with the development of the distinctive spines on their flanks and posterior. Female nymphs shed their skin a further time to reach maturity, while the third instar males enter a resting stage to give rise to winged, short-lived individuals. Like the scale insects, the mealybugs also have a very complex reproductive biology with peculiarities in the way that male and female genes contribute to the genome of the offspring, which is thought

The Important Pest Mealybug Species, Their Host Plants, and Their Geographic Distribution

Species	Host plants	Distribution
Dysmicoccus spp. (pineapple mealybugs)	Pineapple/ sugarcane	Throughout the tropics
Ferrisia virgata (striped mealybug)	Many host plants	Throughout the tropics
Phenacoccus manihoti (cassava mealybug)	Cassava	South America and Africa
Phenacoccus solani (solanum mealybug)	Many host plants	Worldwide
Pseudococcus spp. (tuber mealybugs)	Many host plants	Worldwide
Saccharicoccus sacchari (pink sugarcane mealybug)	Sugarcane	Throughout the tropics
Paracoccus marginatus (papaya mealybug)	Papaya	Throughout the tropics
Pseudococcus elisae (banana mealybug)	Banana	Neotropics and banana-growing regions in the United States

Further Reading: Hill, D. S. *The Economic Importance of Insects.* Chapman & Hall, London, 1997; Capinera, J. L. *Encyclopedia of Entomology,* Vol. 4. Springer, Dordrecht, Germany, 2008; Pimental, D. *Encyclopedia of Pest Management.* CRC Press, Boca Raton, LA, 2002.

to be influenced by the presence of maternally inherited symbiotic bacteria that make it possible for mealybugs to thrive on protein-deficient sap. Impacted by these bacteria, mealy bugs reproduce sexually or asexually and some of the latter are known to produce live young.

Like the scale insects, the mealybugs are pests because they suck sap and produce copious amounts of honeydew that attracts ants and encourages the growth of fungi. Furthermore, they transmit or facilitate the entry of pathogens into their host plants. They are pests primarily in tropical and subtropical regions, although there are a few species known to be glasshouse pests in temperate latitudes. With their soft bodies and lacking the protective scale of their relatives, the mealybugs are primarily found on the aerial parts of the plant, specifically microhabitats that afford them a degree of protection from the drying effects of the sun, such as the underside of leaves, alongside the veins. There are also a few subterranean mealybug species that infest the roots of their host plant.

Mealybugs use a wide range of plants as hosts and they can be pests of many food crops, ornamental plants, and house plants. Some of the more important mealybug pests can be seen in the sidebar. The cassava mealybug (*Phenacoccus manihoti*) is a serious pest of cassava, a very important food crop, especially for subsistence growers in tropical areas. The global production of this crop in 2007 was just over 200 million tonnes, with the biggest producers in sub-Saharan Africa. Cassava is a native of South America and it was introduced to Africa around 300 years ago, where it flourished free from its pests. However, in the early 1970s, the cassava mealybug was accidentally introduced and this pest, free from its own enemies, has gone on to devastate cassava crops throughout Africa. Crop losses caused by this pest can be anywhere between 40 and 80 percent, which can be devastating for subsistence farmers who depend on cassava to feed their families.

Like scale insects, mealybugs can be controlled with cultural, biological, and chemical methods. Plants have a degree of natural resistance to sap-sucking herbivores such as mealybugs, but this resistance depends on the plants having adequate water and nutrients. Nutrient- and drought-stressed plants are more susceptible to becoming severely weakened by plant-feeding insects. In addition to making sure that a plant has sufficient water and nutrients, parts of the plant heavily infested with mealybugs can be pruned.

Biological control has been moderately successful as a means of controlling mealybug pests, especially those that have been introduced to new areas, as was the cassava mealybug. A small parasitic wasp (*Epidinocarsis lopezi*) identified as a predator of cassava mealybug in South America was introduced to Africa in an effort to curb the burgeoning mealybug problem there and to date the introduction has been very successful. Preventing ants from accessing plants can also render the mealybugs more vulnerable to the enemies that are repelled by these pest protectors.

Various insecticides can also be used to control mealybug pests, but in many situations the use of these compounds is beyond the means of poor, subsistence farmers. Insecticides also kill the natural enemies of mealybugs, disturbing the natural mechanisms that regulate pest populations.

FURTHER READING

Hill, D. S. *The Economic Importance of Insects.* Chapman & Hall, London, 1997.

Pimental, D. *Encyclopedia of Pest Management.* CRC Press, Boca Raton, LA, 2002.

Mediterranean Fruit Fly

Fruit fly is the name given to a huge variety of fly species, not all of which are closely related. The adults are generally small to medium sized insects, 5–12 millimeters long, and some of them are capable of damaging a number of food crops. Perhaps the most important fruit fly species from an agricultural perspective is the Mediterranean fruit fly (*Ceratitis capitata*), commonly known as the medfly. Thought to be a native of equatorial Africa, this insect first spread to the Mediterranean region in the 17th century, probably as a result of trade between the countries of Europe and their numerous colonies in Africa. From there, aided by international trade, this little fly has found its way around the world in infested fruit. The mainland United States is currently free of this pest, but in the past it has been recorded in Florida, Texas, and California and subsequently eradicated.

As its name suggests, the Mediterranean fruit fly is a pest because it damages the fruit of a range of plants. At least 250 species of plant are known to be attacked by this insect, but the preferred hosts are apple, apricot, cherry, feijoa, grapefruit, mandarin, orange, passion fruit, peach, pear, persimmon, and plum. Other less important hosts include tomatoes, coffee, peppers, tropical almond, olives, and prickly pear cactus. With such a wide range of plants known to be attacked it is safe to assume that just about any fleshy fruit is vulnerable to medfly damage.

The adult female fly deposits around 300–800 eggs in her 2–3-month lifetime in small groups of 1–14 just beneath the skin of the host plant's fruit. When the larvae hatch they tunnel deeper to feed on the pulp beneath. Depending on the temperature, the larvae complete their development in 7–24 days before tunneling from the fruit to pupate in the soil. The adults of the new generation emerge and disperse to find mates of their own, thus completing the life cycle. The male flies are capable of dispersing around 100–200 meters in search of mates, although greater dispersal (approximately 1 kilometer) has occasionally been seen. In optimal conditions, such as those found in the warm lowlands of Hawaii, the medfly can complete its life cycle in as little as 30 days, but in other cooler parts of its range this can be lengthened up to 100 days. In areas such as Hawaii, the medfly's rapid development allows it to squeeze as many as

A close-up of a Mediterranean fruit fly. (iStockPhoto)

12 generations into a single year. This translates into explosive population growth and it is not unusual for the medfly population in optimal conditions to expand 100-fold each generation.

In heavy infestations the huge numbers of larvae can do a great deal of damage to the fruits of their host plants. The fruit can be damaged to the extent where it drops from the tree or the extensive feeding damage of the maggots makes the fruit unfit for human consumption, mainly because the damage is unsightly and affluent consumers will simply not buy any fruit that looks anything less than perfect. In either case, it is not unusual for 20–50 percent of the crop to be lost due to medfly damage. It has been estimated that the huge fruit-producing industry of California could suffer dreadful losses if the medfly ever became permanently established there. Annual losses could run to $1.8 billion and the jobs of around 14,000 people could be affected if the Medfly ever became established in California. In the nightmare scenario of the pest becoming established across the United States, the annual losses to the fruit industry would be more than $10 billion.

With a huge, lucrative industry at stake it is no surprise that medfly-free countries do their level best to maintain the status quo. Strict quarantine measures are in place to prevent any infested fruit from entering the countries in question and in the event of an infestation the authorities are quick to bring all the means at their disposal to bear on this dipteran interloper. Controlling this fly is complicated by the fact that the larvae hide out for most of their life beneath the skin of their host plant's fruit, enabling the fly to be shipped around inadvertently, especially in the absence of strict quarantine restrictions. When emerging adults betray the extent of an infestation, the standard approach is to spray lots of synthetic insecticides, enforce embargoes on the movement of fruit, and initiate intensive monitoring in the areas surrounding the outbreak.

During the control of medfly outbreaks in California, scientists have also resorted to the use of the sterile insect technique, the conceptually brilliant strategy developed to eradicate the devastating screwworm (see screwworm entry). Conducted swiftly and efficiently these measures can halt the incipient invasion in its tracks, but a tardy or an inappropriate response can allow the flies to get a foothold and they eventually become firmly established, making eradication essentially impossible. To date, the mainland United States, New Zealand, and Chile have managed to eradicate this fly from their borders and the authorities in each country have a steely determination to hang on to their medfly-free status, which they will need. Keeping this small but very expensive fly out of a country is a battle with no end. In the light of climate change and increasing international trade, preventing pests like the medfly from becoming established may become an ever-greater challenge.

FURTHER READING

Lockwood, J. A. *Six-legged Soldiers: Using Insects as Weapons of War.* Oxford University Press, Oxford, United Kingdom, 2009.

McPheron, B. A., and G. J. Steck, eds. *Fruit Fly Pests: A World Assessment of Their Biology and Management.* St. Lucie Press, Delray Beach, FL, 1996.

Mosquitoes

If you didn't know better you would say that the mosquitoes, with their feeble-looking bodies, must be harmless insects, probably content to sit on a flower, sucking nectar. However, as we all know, the truth is very different. The dainty appearance of these flies belies their destructive potential. Mosquitoes are specialized bloodsuckers, perfectly adapted living syringes that have plagued terrestrial vertebrates for millions of years. The oldest fossil mosquito is 90–100 million years old, but it is likely these insects have been buzzing around land-living vertebrates and sucking their blood for at least 150 million years.

Globally, there are around 3,500 species of mosquito and the inherent nuisance quality of most of them doesn't extend beyond the distinctive whine they make as they fly. However, many mosquito species act as vehicles for pathogenic organisms, which cause disease in humans and many of the animals we have domesticated. Technically, these mosquitoes are known as vectors as they transmit pathogens to other organisms. Mosquitoes are vectors for many microorganisms, including the causative agents of malaria, yellow fever, and dengue, to name but a few. The ability to transmit harmful pathogens combined with their abundance and geographic range makes the mosquitoes the most important insect on the planet from a human and animal health perspective. Mosquito-borne diseases have shaped the course of human history and continue to do so today, yet it is difficult to quantify the full impact of these flies on the

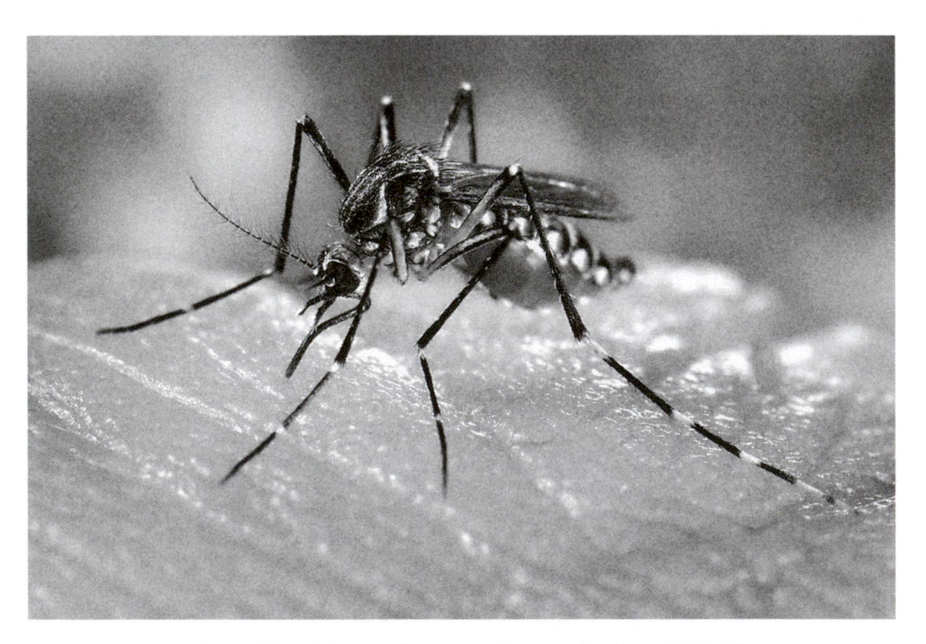

A mosquito sucking blood from a human. (James Gathany/CDC)

human race. There are some statistics that give us an idea of how damaging these insects are, even in the 21st century. For example, it is estimated that each year, mosquito-borne diseases are responsible for the deaths of around 2 million people, although the real number is probably far higher. Humans in affluent, Western cultures rarely feel the effects of these diseases unless they live in semitropical climates or if they choose to holiday in areas where mosquitoes routinely transmit disease to humans. It is the poor, developing countries where the bloodsucking behavior of mosquitoes has the most devastating consequences.

Ever since scientists worked out how mosquitoes transmit diseases, governments and international organizations have pulled out the stops to try and eradicate these flies, but 60 years have elapsed since this war began in earnest and the truth is that we are no closer to controlling mosquitoes than we were in the 1940s. The stark reality is that we may even be losing the fight as insecticide resistance, climate change, human activities, and political and economic factors contribute to an ever-greater geographic range of mosquito abundance.

Mosquitoes may be a bane for humans, but they are very successful animals adapted to exploit the bloodsucking habit to the full. Bloodsucking is the work of female mosquitoes as the nutrients in this liquid are the building blocks for their developing eggs. Just about any vertebrate is fair game for these female flies and some mosquitoes have preferences for a handful of hosts, while other are less selective and will take blood from a range of amphibians, reptiles, birds, and mammals. Male mosquitoes, on the other hand, have no need for lots of proteins and fats, so they feed on nectar. Indeed, male mosquitoes and the females of some species have been shown to pollinate orchids and it is likely they also aid the pollination of many other plants—an often neglected aspect of mosquito ecology. Like many insects, the life span of the adult female can be short—around two weeks—as the sole purpose of this life stage is to reproduce. Once a female's eggs have been fertilized by a male she deposits them in small batches on the surface of water as diminutive rafts or singly in soil. The eggs hatch and the mosquito larvae begin their aquatic existence. Some mosquito larvae are filter feeders, using bristly appendages to strain the water for particles of edible matter, while others browse on the algae and bacteria covering submerged surfaces. There are even predatory species that feed on other small aquatic animals, including other mosquito larvae. These early aquatic stages are one reason why mosquitoes are so successful, for they can complete their development in the most unlikely places, from the stagnant water in an old bucket to the tiny reservoirs that accumulate in tree holes. Just about any standing water can be used as mosquito breeding pools. Even arid environments are not without their mosquitoes as females in these habitats deposit their eggs in the soils and here they wait to be inundated by flood waters.

In the adult mosquito, the normal insect mouthparts have been massively modified via evolution to form a piercing tube, which is inserted through the tough skin of a host and into a suitable capillary. The length of the mosquito's proboscis allows it to penetrate relatively deep capillaries compared with many other bloodsucking flies. Feeding from a large animal can be dangerous, so the mosquito has an array of chemical weapons to help it feed swiftly and painlessly. Compounds in the mosquito's saliva dampen the immune response of the host, reducing the chances of it being alerted to the feeding activity of the fly as well as preventing the blood from clotting in the insect's narrow mouthparts. It's only after the mosquito has finished sucking the host's blood that the feeding site begins to swell and itch, a reaction that can be very severe in some people, while in others there may be only a very weak, barely noticeable reaction. The influx of saliva into the host's body is the route via which pathogens are transmitted by mosquitoes. The types of disease-causing organisms transmitted by mosquitoes can be divided into protozoa, viruses, and nematodes. A veritable library of information has been written on mosquito-borne diseases and it is beyond the scope of this book to look at these conditions in detail, but we will consider the more important protozoan, viral, and nematode diseases of humans below.

PROTOZOAN DISEASES—MALARIA

No mention of mosquitoes as a pest is complete without saying something about malaria. Approximately half of the world's population is at risk from malaria, a disease caused by microscopic protozoan parasites in the genus *Plasmodium*. Every year, malaria infects around 250 million people and leaves around 1 million of these dead—mostly children living in sub-Saharan Africa. The life cycle of this parasite is complex, but one stage invades red blood cells, and then multiplies and digests these cells. In doing so the host's body is flooded with the waste products from the digestion of millions of these cells and the victim's ability to transport oxygen is severely impaired. Often, the havoc wrought by these parasites in the red

Important Mosquito-borne Diseases of Humans

Disease	Disease-causing organism	Distribution
Malaria	Protozoa (*Plasmodium* spp.)	Pantropical
Dengue	Virus	Tropics
Yellow fever	Virus	Africa, South America
Encephalitides	Viruses	Africa, America, Asia, Australasia
Filariases	Nematodes	Pantropical

Adapted from: Hill, D. S. *The Economic Importance of Insects.* Chapman & Hall, London, 1997.

blood cells can be fatal. Even if they survive, people with malaria experience intense fevers and long periods of feeling generally unwell to such an extent that they can't work or look after family.

When numerous communities and whole regions are tormented by this disease, the implications for a developing nation's economy can be disastrous. In countries with a high rate of malaria it has been estimated that economic growth rates can be cut by as much as 1.3 percent, a seemingly trivial figure, but if such economic retardation occurred in developed countries it would be declared a financial disaster.

For several decades a stalemate has existed between us and this parasite, in that a certain degree of control has been made possible with improved knowledge of the parasite's life cycle, antimalarial drugs, and insecticides. Disturbingly, however, it appears the tide may be turning in favor of malaria. Relatively few antimalarial drugs are available and it seems the malaria parasite is developing resistance to even the most potent of these. There are also the looming specters of insecticide resistance and climate change. Mosquitoes the world over are developing resistance to the chemicals that are used to control them and as the climate warms the mosquito vectors of malaria may return to areas where they have not been seen for centuries.

VIRAL DISEASES—DENGUE

Dengue, also known as *epidemic hemorrhagic fever* and the very descriptive *break-bone fever,* is an example of a mosquito-borne virus that is shaping up to be quite a public health problem. Like malaria, mosquitoes pick up the dengue virus by sucking the blood of an infected person, but the virus can also hitch a lift into subsequent generations of mosquito by infecting the eggs of female mosquitoes—so-called vertical transmission.

Normally, the virus causes very severe, flu-like symptoms with extreme pain in the muscles and joints, hence the break-bone moniker. In some cases the virus can cause hemorrhaging of the lungs, digestive tract, and skin. Without treatment, the mortality rate of those infected with dengue can exceed 20 percent. Until fairly recently, dengue was little more than a nasty tropical novelty, but in the last two decades the disease has become one of the big players in tropical medicine (tropical medicine deals with those diseases, often infectious, that are found almost exclusively in the tropics). The World Health Organization estimates that 50 million cases of dengue occur around the world every year. Not only is the disease spreading, but explosive outbreaks are occurring, one of which resulted in 80,000 Venezuelans contracting the disease in 2007, of which more than a quarter of the cases were the hemorrhagic variety.

NEMATODE DISEASES—FILARIASIS

Dangerous microscopic protozoa and viruses are not the only organisms that can be transmitted by mosquitoes. The mosquito menagerie also includes other animals that use humans as hosts. Certain nematodes are transmitted by mosquitoes and these cause a condition called filariasis.

This disease is one of the great neglected areas of public health policy because it is very rarely fatal and it is restricted to tropical and subtropical countries. The fact that filariasis receives little attention does not detract from its ability to debilitate more than 120 million people in at least 83 countries. Like all other parasitic organisms that depend on the mosquito for transmission to other hosts, the life cycle of these nematodes is complex; suffice to say they reach the human host by breaking out of the mosquito's piercing proboscis as it sucks blood. They penetrate the skin via the tiny hole made by the fly's proboscis and enter the lymphatic system, where they mature into adult male and female nematodes. The worms reproduce and their numbers swell to such an extent that they impede the flow of lymph, eventually blocking certain channels completely. These obstructions cause various inflammatory symptoms and even massive swelling of the lower limbs, groin, and genitals, a symptom commonly known as elephantiasis, which is something of a long-established misnomer as it literally means "a condition caused by elephants"!

The nematodes are unique among disease-causing, mosquito-borne organisms because they can be transmitted by at least 77 species and subspecies of mosquito, far more than the protozoa and viruses, which are often transmitted by a single mosquito species or genus. Filariasis is also unique in that it is one of the few mosquito-borne diseases where a real chance of eradication exists. A strategy coordinated between health organizations and pharmaceutical companies has committed large sums of money to the eradication of this disease by 2020, a cooperative effort that meets the objectives of the health organizations and provides the pharmaceutical companies with a lot of good press.

Ever since mosquitoes were identified as the vectors of serious disease in humans and domesticated animals we have sought to control and eradicate them. In the developed world this has been achieved with some considerable success. Until relatively recently, malaria was endemic in parts of the United States and southern Europe, but concerted efforts to kill mosquitoes and modify their breeding habitats in the early and mid-20th century were successful. By 1951 the disease had been effectively eradicated from the United States, and Spain officially declared its malaria-free status in 1964. Controlling mosquitoes can be a slow and very expensive process dependent on killing both adult and larval mosquitoes and modifying the habitats in which the vector species live. Adult and larval mosquitoes can be killed with insecticides or nonselective chemicals and the larvae can also be controlled with the introduction of predatory species, especially certain species of fish (i.e., mosquitofish—*Gambusia affinis*). Wetland areas where the larvae develop can be drained and the number of receptacles and recesses where water can accumulate can be reduced.

Insecticides were once hailed as the nail in the coffin of vector mosquitoes, but it is only in the last 40 years or so that the devastating side effects of these chemicals have become apparent (see introduction). Similarly, wetland drainage schemes can destroy huge areas of pristine habitats that support a wealth of wildlife and the consequences of introducing nonnative species have been seen all over the world. The environmental costs of mosquito control are impossible to ignore, so perhaps the most effective way of limiting the human toll of mosquito-borne diseases is by preventing these insects from biting in the first place with nets and other barriers.

FURTHER READING

Bockarie, M., and D. Molyneux. The end of lymphatic filariasis? *British Medical Journal* 338(2009): 1470–72.

Borkent, A., and D. A. Grimaldi. The earliest fossil mosquito (Diptera: Culicidae), in mid-Cretaceous Burmese amber. *Annals of the Entomological Society of America* 97(5)(2004): 882–88.

Foster, W. A., and E. D. Walker. Mosquitoes. In *Medical and Veterinary Entomology* (G. R. Mullen and L. A. Durden, eds.), pp. 201–48. Academic Press, San Diego, CA, 2009.
Goddard, J. *Infectious Diseases and Arthropods.* Humana Press, Totowa, NJ, 2008.

Pharaoh Ant

This tiny ant has become a real nuisance all around the world and it will come as no surprise to learn that the rise of this insect as a pest has been made possible by the commerce and global travel that have inadvertently transported it to the four corners of the earth. The origins of the pharaoh ant are unclear, but it is generally thought to be a native of Africa. In most places where it has been introduced it seeks refuge from cool temperate conditions by taking up residence in heated buildings, a behavior that has made it one of the most common household ants throughout much of its current geographical range. It is also one of the most difficult household ants to get rid of.

It is the workers of this species that most people are likely to see. They are very small, with a body length of one to two millimeters, so small

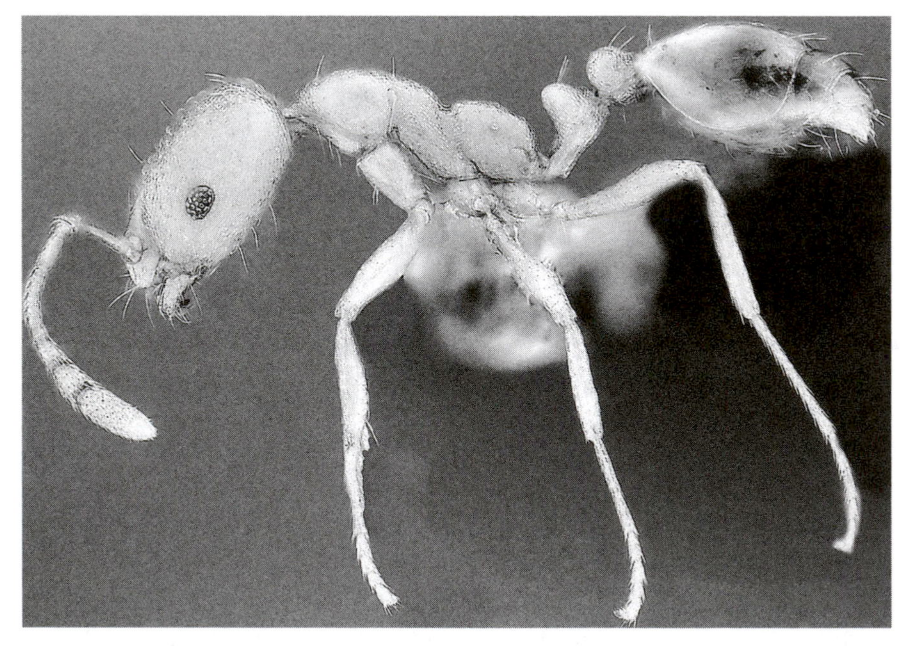

An adult pharoh ant. (U.S. Geological Survey)

they are often overlooked. The queen ant founds her nest in an inaccessible area that is warm (26–30°C) and humid (80% relative humidity), such as wall spaces and other out-of-reach places, often near sources of food and water. Colony size varies from a few dozen individuals to over 2,000. Large infestations are typically caused by many colonies inhabiting the same area. Inside each colony there are several queens (as many as 200 in large nests), males, workers, and all the immature stages (eggs, larvae, pre-pupae, and pupae). In optimal conditions it takes around 38 days for an egg to develop into an adult worker. The queens and males take about 42 days to develop from an egg into an adult and when they're fully developed mating takes place in the nest, which is in contrast to many ant species where the alates leave the nest to join mating swarms of individuals from other nests.

One reason why these ants are so good at colonizing areas into which they have been introduced is that there are many queens in each nest. A colony of pharaoh ants will split, or bud, intermittently. Budding means queens leave the nest, taking some of the workers, and brood with them to found a new nest in an alternative site. Even in the absence of a queen, worker pharaoh ants can stimulate the development of a reproductive female from the existing brood, which allows a small, nascent colony to survive even if the queen dies for some reason. This colony-splitting behavior allows them to colonize a large building in a period of months.

The pharaoh ant is a pest for three main reasons. Firstly, they consume whatever food they can find and their small size allows them to enter containers and packaging to feed on the edible matter within. Secondly, they have a habit of entering and forming colonies in tiny spaces, some of which can cause a potential hazard, such as inside electrical equipment, where they can cause shorting that increases the risk of fires. Even in seemingly hermetically sealed, high-tech laboratories, pharaoh ants somehow find a way in and form colonies. Lastly, their propensity for scuttling around in dirt and decaying matter means they inevitably pick up and disseminate various microbes, some of which are pathogenic. This is a particular concern in hospitals where the abundance of feeding opportunities and nooks and crannies in which to found nests allows large infestations to develop. It has been shown that pharaoh ants can transmit over a dozen pathogenic bacteria, including *Salmonella, Staphylococcus,* and *Streptococcus* species, all of which can cause serious infections in susceptible patients, such as those with open wounds and newborn children. Foraging worker pharaoh ants have been observed in hospitals trying to obtain moisture from the mouths of sleeping babies and bottles supplying intravenous fluids. In hospital situations they also enter high-tech equipment, which causes malfunctions, and they contaminate sterile equipment by simply walking over it.

These ants are so small that an infestation can have already built up to a considerable size by the time it is detected. Worker ants forage around the building searching for suitable sources of food, which can be anything from a small soft-drink spillage to dead insects or silk textiles. When a worker does find a food source it returns to the nest, leaving a pheromone trail that will show others the way, and before long a steady stream of workers scuttles back and forth to the location, collecting food for the nest. The foraging workers will often move between rooms along central heating pipes or electrical cables. It is the presence of the foraging trails that confirms the presence of a pharaoh ant infestation.

Controlling these ants can be very difficult because an infested building is probably home to multiple nests and these nests are typically in inaccessible areas. Sprays and dusts must be avoided because they force the ants to scatter, which can make the infestation much worse in the long term. The tried and tested technique for eradicating pharaoh ants is the use of baits. These can consist of ground-up food laced with boric acid left in suitable containers in close proximity to where the nest(s) may be. The laced food is taken back to the nest where it is consumed by the queen and developing brood, eventually wiping out the colony. Baits can also be laced with compounds known as insect growth regulators, which mimic the effect of natural insect hormones. These chemicals are similarly taken back to the nest; once ingested, they stop the development of the pharaoh ant's brood and prevent the queen from laying any more eggs. Over weeks and months, the nest slowly dwindles away and dies. In any building with a pharaoh ant problem, treatment must be thorough, because any nest that escapes will simply serve as a seed for a new infestation.

Globally, the economic cost of treating pharaoh ant infestations must be large, certainly tens of millions, if not hundreds of millions of dollars every year. Also, as food wastage and a desire for homes with better heating and insulation increases, pests such as the pharaoh ant will undoubtedly become more of a problem. Limiting the spread of the pharaoh ant is ultimately the most effective means of controlling this pest, a strategy that is made more complicated by its small size.

FURTHER READING

Beatson, S. H. Pharaoh ants as pathogen vectors in hospitals. *Lancet* 1(1972): 425–27.

Hoelldobler, B., and E. O. Wilson. *The Ants.* Belknap Press, Cambridge, MA, 1990.

Planthoppers

Planthoppers are small, true bugs in the family delphacidae and are characterized by mouthparts that are modified to form a rostrum used to pierce the outer tissues of their host plants to get at the sap-containing phloem vessels within. As their common name suggests, planthoppers are also accomplished jumpers, using their hind legs to propel them into the air at the first sign of danger. Planthoppers are found around the world. Currently, around 2,000 species are known and it is highly likely that many

The brown rice planthopper can be a very serious pest of the world's most important food crop. (Nigel Cattlin / Visuals Unlimited, Inc.)

more species remain to be identified, especially in the humid tropics where they are at their most diverse.

Planthoppers begin life as eggs deposited in fissures on the surface of their host plant or even beneath the bark. Nymphs hatch from these eggs and begin to feed from the host plant by drawing sap through their long, hollow mouthparts. In some species, the planthopper nymphs produce a waxy secretion like the mealybugs and scale insects that probably serves to protect them from desiccation and their many predators. The nymphs go through a number of instars, shedding their skins at the end of each one to allow them to grow. Eventually, after three of four instars, the planthoppers reach maturity and find themselves equipped with fully functioning wings and gonads. The insects use their prodigious jumping abilities to take flight from their predators and also as a way of taking to the air without laboring their wing-muscles. Planthoppers are among the most accomplished jumpers in the animal kingdom, a feat made possible by the presence of the rubbery protein, resilin, which is compressed by the action of the hind leg being cocked ready for takeoff. The catch mechanism is released and the insect is launched explosively into the air with an acceleration of 700g (units of gravitational force exerted on the body), a force that would tear a human limb from limb.

With their wings and resilin-powered jumping abilities, the planthoppers are in no way sedentary like their relatives, the scale insects. They can take to the air at will to search for new food plants and mates, an ability which makes them very troublesome pests. Only a very small minority of all the planthopper species known to science are considered pests. They are of primary importance as pests of cereal crops, especially rice as well as sugar cane. In some areas of the world they are considered to be the number one insect pest of rice. Their sapsucking drains the plant of valuable nutrients and they are also known to act as vectors for simple, albeit very damaging plant bacteria known as phytoplasmas. These bacteria undergo some of their development in the planthopper host and are transmitted to the plant when the insect feeds. Inside the plant, these bacteria are obligate parasites, feeding on the cells of the phloem vessels. The symptoms of phytoplasma infection range from slight yellowing of the leaves to death of the infected plant, underlining their importance in agricultural systems. The planthoppers are also known to transmit viruses from plant to plant, many of which can be very destructive in commercial crop cultivation. As with all sap-feeding bugs, the planthoppers excrete the excess water from the liquid food they imbibe as honeydew, which serves to encourage the

The Important Pest Planthopper Species, the Crops They Attack, and Their Geographic Distribution

Species	Host plants	Distribution
Laodelphax striatella (small brown planthopper)	Cereals and sugarcane	The northern hemisphere
Nilaparta lugens (brown rice planthopper)	Rice	Southeast Asia, from India to China
Peregrinus maidis (corn planthopper)	Maize, sorghum, and sugarcane	Throughout the tropics
Perkinsiella saccaricida (sugarcane planthopper)	Sugarcane	Australia and Hawaii
Sogatella furcifera (white-backed planthopper)	Rice	Southeast Asia

growth of fungi and to attract ants, both of which are ultimately detrimental to plant health.

Planthopper pests are also associated with the phenomenon known as hopperburn, where the leaves of the infested plant turn brown and wilt. It is heavy infestations of planthopper that cause hopperburn and it was once thought to be caused by toxic compounds in the saliva of the planthopper. It has since been discovered that hopperburn is actually a plant response to wounding, triggered by the way in which the planthopper's mouthparts move in the plant, and exacerbated by the insect's saliva.

The most important planthopper pest is the brown rice planthopper, *Nilaparvata lugens,* which feeds on the sap of this very important crop. In 2007, the global rice crop was more than 600 million tonnes. For a good proportion of the human population, rice is their staple diet. Any pest that causes significant losses to the rice harvest can directly affect the lives of many millions of people. Very conservative estimates from the late 1970s of the annual economic losses inflicted by the brown rice planthopper on the rice-growing industry were in the region of $300 million. Today, we can be certain this figure is many times higher, probably several billion dollars per year. Until recently, this planthopper species was only considered to be a serious pest in Japan. However, during the 1970s, it became a much more serious pest throughout Southeast Asia and today it is one of the most serious insect pests of rice in the major rice-growing regions of the world.

The brown rice leafhopper is a pest not only because of the sap it sucks from its host plant, but also because it transmits the potentially devastating viral disease known as rice grassy stunt virus. Rice varieties vary considerably in their resistance to this virus, but some of the more high-yield varieties favored by many farmers are acutely susceptible to this virus. The brown rice leafhopper has a number of characteristics that predispose it to becoming a pest, especially in modern rice cultivation systems. Firstly, this insect has good powers of dispersal, meaning that it can find its way to new rice-growing areas. Secondly, once it has reached a new patch of habitat, its high level of fecundity means that its populations can quickly reach levels that surpass economic thresholds. Thirdly, it seems the application of nitrogenous fertilizers to the rice crop can actually improve the insect's tolerance to adverse environmental stresses.

Planthoppers can often be difficult to control in commercial agricultural systems. As with many pests the inability to control a problematic planthopper species follows logically upon the way in which commercial agriculture operates: the steady erosion of biodiversity creates an environment where the populations of a pest's natural enemies are reduced to such an extent that they can no longer provide natural regulation of the pest. In some areas, agriculturalists are realizing that long-term commercial success in growing crops depends on restoring the elements of a natural ecosystem that intensive agriculture stripped away. Limiting the damage caused by insects such as the planthoppers depends on employing cultural methods and biological control, for example, enhancing the populations of predators, parasites, and pathogens by introducing them and making the agricultural environment more conducive to their survival. Chemical control can be useful in some circumstances, but these compounds must be applied carefully and with a full understanding of their environmental impact.

FURTHER READING

Denno, R. F., and T. J. Perfect. *Planthoppers: Their Ecology and Management.* Chapman & Hall, London, 1994.

Pubic Louse

Pubic lice (*Pthirus pubis*), commonly known as crab lice, or simply as crabs, are another species of sucking louse that make their living on humans. As their name suggests, these insects prefer pubic hair, so they are often found in the pubic region and the armpits, but they are also an occasional occupant of beards, moustaches, eyebrows, and eyelashes. It takes a microscope to appreciate the bizarre appearance of these tiny animals. Adult pubic lice are 1.5–2 millimeters long and almost as broad as they are long, which together with their huge, grasping claws gives them a rather crab-like appearance. The big claws are perfectly suited to grasping the relatively thick pubic hair, affording them a good grip on their active hosts.

Compared to the head louse and body louse, the evolutionary history of the pubic louse is a little more complicated, because it appears that the ancestor of this species was a parasite of gorillas. Our ancestors could have

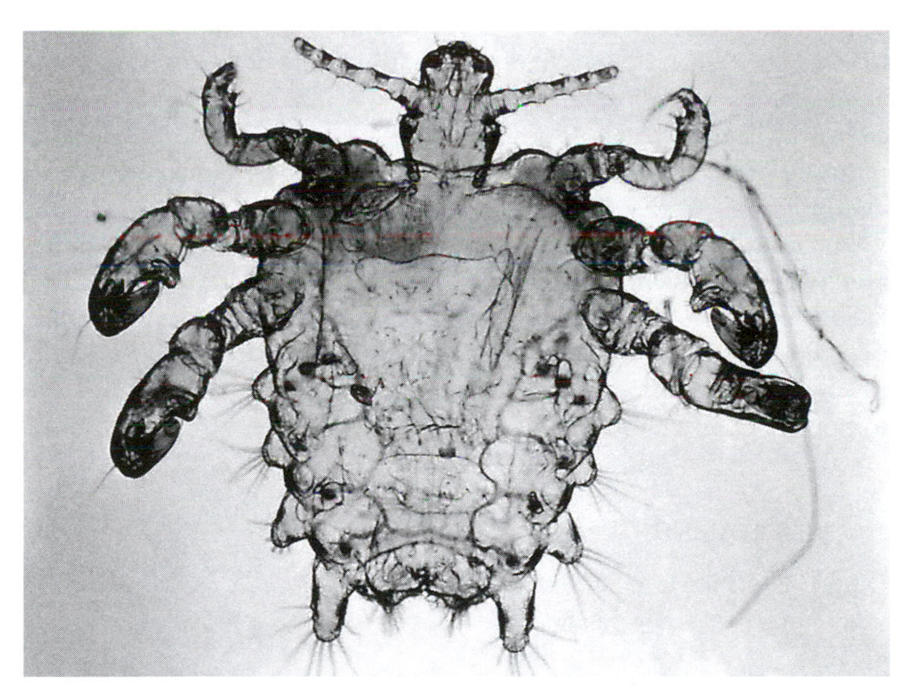

Phthirus pubis, more commonly known as the pubic or crab louse, viewed through a microscope. (WHO/CDC)

picked up this parasite in a number of ways, some of which are rather unsavory. The ancestors of the human pubic louse may have switched hosts when our ancestors hunted gorilla and butchered the carcasses or made use of areas regularly frequented by gorillas. However the association between this louse and humans developed, DNA analysis suggests that the ancestors of this insect started living on our ancient ancestors three to four million years ago.

The pubic louse is far less active than the other parasitic lice of humans and for much of its life it remains in the same place with its mouthparts firmly fixed in the skin. These blood meals are crucial for the maturation of the female's eggs; in her lifetime she will produce around 30 eggs—a fraction of those produced by the head louse and body louse. The eggs are individually attached to hairs and they hatch after seven to eight days. The entire life cycle can be completed in a little over a month.

Crab lice are pests because of the itching caused by the bites and the social ignominy of an infestation. Crab lice do not transmit any pathogens, but their bloodsucking activities can cause intense irritation and the need to scratch the infected areas and seek treatment can be very embarrassing for people who play host to these animals. Crab lice can pass from person to person via bedding and clothing in crowded situations, but the typical means of dispersal to new hosts is via sexual intercourse, making them a venereal problem.

Pubic lice are a problem the world over. Whether among the populations of the affluent nations or in areas where education and medical facilities are poor, these insects may parasitize 10 percent of the population. Preventing and treating crab lice is not complicated and it typically involves the application of insecticides to infected areas, improvements in hygiene, and raising awareness of the risks of casual sex.

FURTHER READING

Durden, L. A., and J. A. Lloyd. Lice (Pthiraptera). In *Medical and Veterinary Entomology* (G. R. Mullen and L. A. Durden, eds.), pp. 56–80. Academic Press, San Diego, CA, 2009.

Reed, D. L., J. E. Light, J. M. Allen, and J. J. Kirchman. Pair of lice lost or parasites regained: The evolutionary history of anthropoid primate lice. *BMC Biology* 5(7)(2007).

Red Imported Fire Ant

Like the cane toad and the European rabbit, the red imported fire ant, *Solenopsis invicta,* has become a problem simply because it has been introduced by humans into areas beyond its native range—the Mato Grosso of Brazil. It was first introduced into the United States through the ports of Mobile, Alabama, or Pensacola, Florida, at some point between 1933 and 1945, the exact details of which are unknown. It was transported via ship, perhaps with the soil around crops or with other goods. The original founder population was no more than 9–20 mated females, each of which had the potential to start a nest. Apart from this initial introduction, it is very likely other batches of ants were also inadvertently introduced at later dates. These founder populations quickly grew and today the introduced ant is found in Alabama, Arkansas, California, Florida, Georgia, Louisiana, Maryland, Mississippi, New Mexico, North Carolina, South Carolina, Oklahoma, Tennessee, Texas, and Virginia. In addition, it has also been introduced into Antigua and Barbuda, the Bahamas, Puerto Rico, the British and U.S. Virgin Islands, Cayman Islands, Hong Kong, Malaysia, Singapore, Taiwan, Southern China, Philippines, Trinidad and Tobago, the Turks and Caicos Islands, Australia, and New Zealand.

Like all ants, the red imported fire ant has a very interesting natural history. All the ants in the nest, regardless of their caste, are the offspring of the queen, making them siblings. It is this relatedness and seamless cooperation that makes ant colonies so interesting biologically. Everything an ant does, all its frantic foraging and dogged tenacity in defending its home turf, is for the good of the colony. Ecologically, an ant colony made up of tens, hundreds, or thousands of individuals functions as a single organism—a superorganism. Red imported fire ant nests are subterranean, but a mound, rarely more than 45 centimeters across, is visible above the ground. The mound is occupied by the queen and a raft of workers, including small, medium, and large individuals, all of whom have slightly different tasks in the colony. When food is plentiful the queen can produce 1,500 eggs a day and with a life span of two to six years she can produce an impressively large number of workers. An egg takes between 22 and 38 days to develop into an adult worker with a life span anywhere between 30 and 180 days, depending on its size and function. Each nest is founded by a single female or sometimes a group of cooperating females,

Fire ants have become quite a problem in the areas where they have been accidentally introduced. (iStockPhoto)

each replete with all the sperm they will ever need, obtained during their mating flight from the nest in which they developed. The males die soon after mating, but the females live on, intent on finding a suitable retreat in the ground in which to excavate a small chamber—the nascent nest.

Here the young queens lay their first eggs and feed their first daughters on fats (some of which are sourced by the breakdown of the queens' wing muscles, which are now surplus to requirements), spare eggs, and secretions from their salivary glands. On these slim pickings the first brood of workers is a decidedly poor collection of runts. Regardless of their diminutive stature, the first workers have an innate understanding of their tasks, and guided by pheromones produced by the queen, they begin finding food for the colony and enlarging the nest. Egg laying continues and the nest increases in size, eventually developing into a veritable insect metropolis, often home to more than 240,000 workers frantically going about their business. After about a year, the queen starts producing eggs destined to become winged females and males, known as alates. It is these winged ants that will fly off, mate, and attempt to establish nests of their own, thus completing the cycle.

In its native range the red imported fire ant has to contend with a panoply of predators, parasites, and pathogens, but in the United States and in other areas where it's been introduced the ant has virtually no

Biological Control Agents for Red Imported Fire Ants

Species	Type of organism	What they do	Status
Thelohania solenopsae	Microsporidian (microscopic fungi-like organism)	Infect workers and queen, weakening and reducing the size of the colony	Experimentally released
Beauveria bassiana	Fungi	Infects and kills ants at various stages of growth	Experimentally released
Pseudacteon tricuspis	Fly	Larvae are internal parasites of workers	Experimentally released
Pseudacteon curvatus	Fly	Larvae are internal parasites of workers	Experimentally released
Solenopsis daguerri	Ant	A parasitic ant species that invades and takes over the fire ant nest	Still being assessed

natural enemies and its populations can expand unchecked. When there are large quantities of these ants they can become a problem: damaging crops, preying on native wildlife, and harming people and property. The workers often search for food in food crops, including soybean, citrus, corn, okra, bean, cabbage, cucumber, eggplant, potato, sweet potato, peanut, sorghum, and sunflower. They damage these crops by feeding on the young, tender growth, and it has been estimated the economic toll on the soybean industry alone is in the region of $150 million dollars every year. In urban areas their nest building can damage buildings, roads, and sidewalks. They are attracted to anything with an electrical current, so they often aggregate and even build nests in electrical appliances and junction boxes. They often cause electrical shorts in these devices, resulting in damage and even fires.

Fire ants are omnivores, but the workers defend their nests tenaciously and will attack any interlopers, even humans, by clinging on with their legs and mandibles and plunging their sting into the enemy. For small animals, considerable doses of fire ant venom injected by a large number of these insects can be fatal. Human victims of fire ant attacks can experience localized pain, swelling, and even anaphylactic shock in rare cases. The name *fire ant* actually relates to the pain of the insect's sting; the author has vivid memories of accidentally stepping in a small fire ant nest in bare feet and dancing around to get rid of the stinging workers. Since the red imported fire ant has become a serious problem in the United States, money has been spent trying to stem its expansion across the country. To date, state and federal agencies have spent more than $250 million. Every year, companies and individuals throw $25–40 million at fire ant control and eradication, but because this insect is now so well established, eradication is a practical impossibility and the expense of control will continue to mount.

Insecticides are the foundation of fire ant control and they are applied to nests in various ways, such as drenching, dusting, mound injections and baits, and broadcast spraying, all of which can be locally successful. However, the sheer scale of the fire ant problem is far beyond the practical and environmental limitations of insecticide applications. In recent years, biological control of the fire ant has received a lot of attention. In order to identify potentially useful biological control agents, biologists were charged with the task of journeying to the natural home of the fire ant—the interior of Brazil—where they could observe the species in its native habitat and learn more about its natural enemies. To date, five promising organisms have been identified (see sidebar).

The organisms in the sidebar punish the fire ants in some very grisly ways. The parasitic flies lay their eggs on the worker ants when they're out foraging. The maggot hatches and burrows into the ant, eventually taking up residence in the head capsule, where it eats all the contents. The ant's hollow head falls off and the maggot completes its development, pupates into an adult fly, and begins the cycle all over again. The spores of the fungus infect the ant and grow throughout the body of the insect, eventually bursting out of the hapless victim in a cloud of spores that will infect more ants. As you can see, the fire ant has some very unpleasant enemies and these are now poised to aid in the struggle to control this alien insect.

FURTHER READING

Tschinkel, W. R. *The Fire Ants.* Harvard University Press, Cambridge, MA, 2006.

Rice, Maize, and Granary Weevils

Collectively, seeds and grains of various cultivated grasses are the most important food crops in the world, with hundreds of millions of tonnes being produced each year. Rice, maize, and wheat rate as the most important of the large variety of seed and grain crops; they are processed in myriad ways to feed billions of people around the globe. Long before these plants were domesticated by humans they had their attendant herbivores, some of which were generalists, while others were specialists adapted to feed on specific parts of their host plant. The seeds of these grasses are packets of energy-rich food and certain insects forged an existence exploiting this resource. By cultivating these plants, we provided these specialist seed-feeders with an abundance of food and they have followed the trail of agriculture around the world.

The most important of these seed feeders are the rice, maize, and granary weevils: tiny, elongate weevils belonging in the genus *Sitophilus.* Barely five millimeters long, these beetles are the most serious pests of stored grains and seeds on the planet. The rice weevil (*S. oryza*) is predominantly a pest of rice in warm regions and temperature-controlled warehouses in temperate regions; the maize weevil (*S. zeamais*) is primarily a maize specialist. The granary weevil (*S. granarius*) is primarily a pest of wheat and the grains of related grasses and is arguably the most problematic species of the three in temperate regions. Until quite recently, these weevils were all thought to be morphological variations of the same species, but they are now considered to be three separate species. The adults are two to five millimeters long and bear the curved snout-like rostrum characteristic of the weevil family.

Rice weevils can inflict heavy damage on stored rice. (Liewwk | Dreamstime.com)

The life of these weevils begins with the adult female using the powerful mandibles at the tip of her rostrum to chew a hole into a grain or seed. She deposits one or two eggs, depending on the species, into this cavity and fills the hole with a gelatinous secretion. The first instar larva, a tiny, white and legless grub, hatches from the egg to find itself surrounded by all the food it will need to complete its larval development. In ideal conditions development is fast, taking as little as 26 days from egg to adult. Pupation also occurs in the brood seed or grain and the newly emerged adult has to chew itself free to continue the cycle by dispersing and searching for a mate. Of these three beetle species only the rice weevil has the power of flight—the other two species have secondarily lost this ability.

In her lifetime, an adult female *S. granarius* can produce around 150 eggs and in temperate latitudes there may be as many as three or four generations in a single year. Therefore, a single female grain weevil is hypothetically capable of producing around five million progeny by the third generation. With each developing larva requiring an entire grain or seed in which to complete its development it is not difficult to see how an infestation of these weevils can have a huge impact on stored seeds and grains even in the absence of perfect conditions. Crop losses due to the feeding activity of these weevils can be anywhere between 20 and 100 percent. However, it is essentially impossible to provide an accurate estimate of the economic losses caused by these beetles because in many areas where their host plants are grown and stored there aren't detailed records of crop yields and storage damage from one year to the next. The small size of these weevils and the cryptic nature of the immature stages, that is, the fact that the egg, larva, and pupa are secreted within a single grain, make it difficult to detect an infestation, especially in its early stages. Once an infestation has been detected, a considerable proportion of the entire crop may already have been lost. The first signs of an infestation are warm, moist areas within the stored grain signaling the growth of fungi on the various wastes produced by the developing larvae nibbling away in their individual grains. The level of moisture in the stored crop can become high enough for the seeds to sprout.

All of the *Sitophilus* weevils can exact heavy losses on stored seeds and grain. The most damaging of the three species is the rice weevil, for two reasons: rice is the most important food crop in the world, and the weevil retains the power of flight, allowing the adults to disperse from the grain stores in which they developed to new ones and initiate new infestations.

Controlling the rice, maize, and grain weevils is far from easy and has been complicated in recent decades by the emergence of insecticide resistance. The cheapest way both economically and environmentally to control these beetles is ensuring that any areas where grains and seeds are stored are free from the remnants of the previous year's crop and that any spillages in and around the storage areas are cleaned up so as not to present the weevils with easy pickings from which to colonize the main stores. The containers in which the crop is to be stored should be structurally sound, providing no obvious points of entry for mature beetles looking for a place to breed. The stores should be inspected intermittently to check for any areas of heat or mold betraying the presence of the unassuming beetle larvae feeding within their grains. Should an infestation be discovered, the beetles can be killed by fumigation with various insecticides although this has implications for the future use of the crop as well as the wider environment. Another possibility is increasing the temperature and lowering the humidity in the stores to create conditions far from conducive for the survival of the developing larvae. Other environmentally sound means of controlling these beetles include the use of pheromones to lure males into traps where they can be collected and disposed of, and sticky traps where the flying or walking adults are snared in a very tacky, slow-drying adhesive.

When these beetles first took up with humans millennia ago they were not alone. Like all insects they have their very own band of predators and parasites, some of which can be harnessed by farmers to control infestations of these weevils. Parasitoid wasps, nematodes, and fungi can be very effective in controlling infestations of these weevils because the problem is confined to a closed environment. Unlike biological control in field crops where many of the released natural enemies can simple wander off without doing as they are required, the predators and parasites introduced into a grain silo don't really have anywhere to go, so they get to work eliminating a high proportion of their quarry.

Collectively these weevils are a problem around the world, but it is the people and communities of developing regions who are hardest hit by these pests. Many of the control measures described above are well beyond the simple means of people with very little money and in most cases the infrastructure and equipment needed to store a crop are simply not available. The large, sophisticated, climate-controlled grain silos you see on farms in the developed world do not come cheaply, so poor farmers in developing countries have to keep their harvest in rudimentary storage facilities, which are inadequate for keeping insect pests at bay.

FURTHER READING

Capinera, J. L. *Encyclopedia of Entomology*, Vol. 2. Springer, Dordrecht, Germany, 2008.

Hill, D. S. *The Economic Importance of Insects.* Chapman & Hall, London, 1997.

Pimental, D. *Encyclopedia of Pest Management.* CRC Press, Boca Raton, LA, 2002.

Sand Flies

Sand flies are small delicate insects, no more than about three millimeters long; however, their fragility belies the impact they have on human populations. Sand flies are a serious economic and medical pest because they are the main vectors for the protozoa that cause the various forms of leishmania, a debilitating, often fatal group of diseases, as well as a small number of bacterial and viral diseases.

Around 700 species of sand fly are distributed throughout the tropics and subtropics. Approximately a tenth of these species are thought to be of public health importance. Female sand flies feed on blood, plant juices, and the honeydew secreted by sapsucking insects. The males, on the other

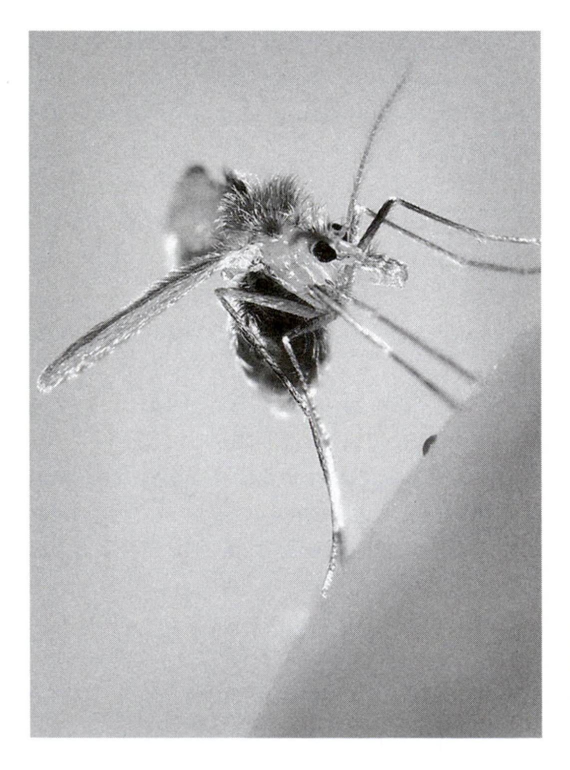

Sandflies are the vector for the protozoan that causes the serious disease leishmania. (CDC/ Frank Collins)

hand, never feed on blood. It is only the female fly that requires the protein contained in blood to complete the maturation of her developing eggs and only she has the mandibles necessary to cut the tough skin of the host to access the liquids below. Depending on the species in question, sand flies take blood from reptiles, amphibians, birds, and mammals, including humans. The soft bodies of sand flies dry out quickly, so most species shy away from the sun, preferring to feed at night or during twilight and early morning when it's cool and humid. Sand flies are also terribly weak fliers, so they are effectively grounded in windy conditions.

Mated female sand flies with access to blood will begin to seek out habitats to nurture their offspring. Typically these are dark places with high humidity and an abundance of organic debris on which the larvae will feed. Suitable larval habitat includes the burrows of animals, crevices, and rot holes in trees and among leaf litter. Depending on the sand fly species, larval development can take between 2 and 10 weeks, with pupation completed in around 10 days.

Of the diseases transmitted by sand flies, by far the most important, medically and economically, is leishmania, an array of very complicated conditions caused by a number of protozoan species in the genus *Leishmania*. The protozoa are picked up by the sand fly from an infected human along with its blood meal and they undergo some of their development in the insect before finding their way into another human when the sand fly next feeds. Like so many tropical and subtropical infectious diseases, leishmania is one of the great neglected areas of public health as it's a major problem in developing countries rather than the affluent developed world. A huge amount has been written on this disease, so we'll have a brief look at what it is and how it affects the human race.

The disease is currently endemic in 88 countries and every year there are thought to be two million new cases. A 10-year epidemic in the Sudan between 1984 and 1994 is thought to have killed 100,000 people from a population of 300,000 in the western upper Nile region of the country. Twelve million people around the world are thought to be infected with the disease, which, depending on the *Leishmania* protozoan species present, manifests in one of three important forms—cutaneous leishmania, mucocutaneous leishmania, and visceral leishmania. All three of these are problematic, but the latter is by far the most lethal of the three. Cutaneous leishmania causes spreading ulcers that form at the site of the sand fly bite, which can lead to secondary infections, extensive scarring, and even death if the secondary infections are serious. Mucocutaneous leishmania begins in a similar way to the cutaneous form, but a

The Five Most Important *Leishmania* Species

***Leishmania* species**	**Distribution**	**Classification of disease**	**Disease names**
L. tropica	Middle East, India	Cutaneous	Cutaneous leishmania, Jericho boil, Aleppo boil, Delhi boil, Baghdad ulcer, Bouton d'Orient
L. major	Africa, Middle East, Asia Minor	Cutaneous	Cutaneous leishmania, Jericho boil, Aleppo boil, Delhi boil, Baghdad ulcer, Bouton d'Orient
L. donovani	Occurs on all contents. Most common in Bangladesh, Brazil, India, Nepal, Sudan, Ethiopia and China	Visceral	Dum-Dum fever, kala-azar, black fever
L. braziliensis	Central and South America: Mexico to Argentina	Cutaneous	Mucocutaneous leishmania, espundia, uta
L. mexicana	Northern Central America, Mexico, Texas, Dominican Republic, and Trinidad	Cutaneous	Chiclero ulcer, bay sore

Further Reading: Rutledge, L. C., and R. K. Gupta. Moth flies and sand flies *(Psychodidae).* In *Medical and Veterinary Entomology* (G. R. Mullen and L. A. Durden, eds.), pp. 147–63. Academic Press, San Diego, CA, 2009; Goddard, J. *Infectious Diseases and Arthropods.* Humana Press, Totowa, NJ, 2008; Lehane, M. J. *The Biology of Blood-Sucking in Insects.* Cambridge University Press, Cambridge, MA, 2005.

secondary lesion develops after the first lesion has healed or as long as 30 years later. This secondary lesion is very nasty: the soft tissues and cartilage of the nose and mouth are destroyed, leading to very unsightly disfiguring, collectively known as espundia. Visceral leishmania usually begins with fever-like symptoms, but can then progress to wasting and anemia followed by enlargement of the liver and spleen and finally death in untreated cases.

Recently, a very worrying synergy has developed between leishmania and HIV, one that has seen health organizations initiate a global surveillance program to monitor this ominous trend. Leishmania infections hasten the onset of AIDS by exacerbating the immune-suppression caused by the HIV virus. In areas where cases of this co-infection are increasingly common, leishmania becomes an epidemic rather than a sporadic threat. In southern Europe, 70 percent of visceral leishmania cases are associated with HIV infection and users of injected drugs are the most seriously affected group.

Treating leishmania is very difficult indeed as the causative protozoa are cunning adversaries. They hide out in the digestion compartments of macrophage cells, the very cells that envelope and kill invading organisms as part of the immune response. Older treatments for the disease were nothing more than poisons based on the element antimony. Newer treatments are still based on antimony, but they are a little safer. As nasty as these drugs are, the *Leishmania* parasites are evolving resistance to them. Vaccination may be possible against some of the *Leishmania* species and research in this area is ongoing.

The other pathogens transmitted by the humble little sand fly include the bacteria that causes Carrión's disease (*Bartonella bacilliformis*), which manifests as a nonfatal cutaneous form (verruga peruana) and a potentially fatal visceral form (Oroya fever). Sand flies are also the vector of a virus in the genus *Phlebovirus,* an organism that causes sand fly fever, also known as pappataci (papatasi) and three-day fever, which as the name suggests is a fever. Although sand fly fever is nonfatal, recovery can take a long time.

The key in controlling these diseases is curbing the populations of sand fly and preventing the adult flies from biting, both of which are easier said than done. Insecticides are routinely used to kill the adult flies clothes and fly nets impregnated with insecticides and insect repellants will prevent bites. Control of the larvae is very difficult as the biology of the immature stages of sand flies is poorly understood, largely due to the difficulty in correctly identifying the larval habitat. Many species of mammal are known to be reservoirs for the pathogens transmitted by sand flies, including dogs and various rodents. Controlling the populations of these animals around human dwellings and even treating dog collars with insecticide has some impact on limiting the transmission of the sand fly–borne diseases.

FURTHER READING

Goddard, J. *Infectious Diseases and Arthropods*. Humana Press, Totowa, NJ, 2008.

Lehane, M. J. *The Biology of Blood-Sucking in Insects*. Cambridge University Press, Cambridge, MA, 2005.

Rutledge, L. C., and R. K. Gupta. Moth flies and sand flies (*Psychodidae*). In *Medical and Veterinary Entomology* (G. R. Mullen and L. A. Durden, eds.), pp. 147–63. Academic Press, San Diego, CA, 2009.

Scale Insects

Insects in the family coccidae, commonly known as scale insects, are extremely odd in appearance and nature—probably among the most peculiar of all the insects. Most people would have trouble identifying them as animals at all, let alone insects. Taxonomically, they are true bugs (hemiptera), quite closely related to aphids, pysllids (plant-lice), and whiteflies. Like all hemiptera, their mouthparts are modified into a long, thin rostrum they use to pierce plant tissues to get at the fluids coursing through the phloem vessels. Globally, around 7,300 species of scale insect have been identified, but many more remain to be identified, especially in the humid tropics.

Immature scale insects (nymphs) are mobile animals that crawl sluggishly on their host plants, but when they mature they become sedentary, remaining in one place with their mouthparts firmly lodged in their host plant. To give them a degree of protection against their many enemies, scale insects secrete a waxy shelter that resembles the scale of a fish or reptile, hence their common name.

The scale insects are a diverse group of animals and they feed on a huge variety of plant species. As a group, they are considered to be one of the most devastating pests of woody plants in the world. Females are mobile in their immature stages (nymphs) and can be found crawling around on the host plant searching for a suitable site to feed and mature. Even into maturity, they retain many of the features of the nymph, whereas the very short-lived males are fully winged, enabling them to fly between colonies to maintain variability in the gene pool. In addition to the morphological variation between the sexes, the scale insects as a group are also known for the diversity of reproductive systems they display. There are sexually reproducing species, parthenogenetic species, and hermaphroditic species.

Adult scale insect sucking sap from a birch sapling. A mass of white, fluffy wax protects the soft scale insect. Like aphids, scale insects are tended and protected by ants, which covet the honeydew these sap suckers secrete. (Courtesy of Ross Piper)

There are even some scale insects where the female has a placenta-like structure to nurture her developing nymphs. For reasons that are still to be understood, when a male and female scale insect mate, the resultant offspring do not always get 50 percent of their DNA from each parent, like in most animals. In scale insects, inheritance of genetic material is often very complex, involving unequal contributions from males and females and deactivation or complete elimination of some or all of the genetic material from the male. Often, a male scale insect is the product of parthenogenetic reproduction—the female can produce sons in the absence of a father. The intricacies of scale insect reproduction are thought to be a result of the symbiotic bacteria that live inside every one of them.

Scale insects are broadly classified into two groups: soft scales and armored scales. Soft scale insects produce a scale that is integral to their body and they are three to five millimeters in size. Armored scale insects, on the other hand, produce a scale that is not an integral part of their body. It serves as a separate cover under which the insect hides and feeds.

In the armored scales, the life cycle begins with the adult female laying all of her eggs beneath her protective scale. Depending on the species, the eggs hatch after one to three weeks and the nymphs crawl out from beneath their mother's scale and seek succulent new growth on the host plant. When they have found a suitable part of the plant they use their mouthparts to penetrate the phloem vessels and suck the sugary fluid within. Now, firmly attached to their host plant, the female armored scale insects shed their skin and lose their legs and antennae as they no longer have any need for them. They shed their skin a second time before reaching maturity. The skins the female sheds are incorporated into her growing scale. The male armored scale insects go through quite a different development process by shedding their skins two additional times and entering a resting stage beneath their waxy shelter. The male armored scale insects normally have one pair of wings and they only live for a day or two, so with no time to lose they take to the air in order to find a female to mate with.

In the soft scale insects, the female's legs and antennae are reduced, but not lost completely, so they can move about, although they seldom do. The females secrete wax that forms a fluffy white sac at the end of their body, and it is into this waxy, fluffy mass the eggs are deposited. The rest of the soft scale insect's life cycle is rather similar to their armored relatives.

Every scale insect, whether it's an armored scale or a soft scale, is an obligate sap feeder. Sap is a food source used by many animals. It is abundant and very rich in sugars, but animals that feed solely on this fluid are faced by a big problem—sap contains only vanishingly small quantities of amino acids, which are crucial for growth and development. To solve this problem, many sap-feeding insects have formed remarkably close relationships with various micro-organisms, typically bacteria that dwell in cells of a specialized structure called the bacteriome. These bacteria feed on some of the sap ingested by the scale insect and produce amino acids, some of which they share with their host. The bacteria are passed from the mother scale insect to her developing young, whereas the bacteria in the male's bacteriome are an evolutionary dead end. The presence of these bacteria and the way they are inherited is one explanation for the myriad peculiarities of scale insect reproduction. It seems the bacteria are doing their level best to eliminate male scale insects because they serve no purpose in the passage of the bacteria from one generation to the next.

Bizarre life histories aside, the scale insects are very important pests of many species of woody plant, including fruit and nut trees, ornamentals, forest plants, greenhouse plants, and house plants. They damage the plants they feed on in a number of ways. Firstly, in heavy infestations, the removal of large amounts of sap can reduce the plant's ability to develop its leaves and fruits fully. Secondly, the sheer volume of sap a scale insect imbibes would quickly inflate it to bursting, but an efficient fluid balance mechanism—the insect equivalent of vertebrate kidneys—quickly extracts the surplus fluid from the sap and excretes it as honeydew, the sugary water that all sap-feeding insects excrete in copious quantities. In heavy infestations of scale insects, the honeydew may be produced in sufficient quantities to coat the leaves and other parts of the plant, impairing the plant's ability to photosynthesize and encouraging the growth of fungi. The production of honeydew is also a problem because it attracts ants, the natural bodyguards of scale insects. Ants relish honeydew and to protect their supply of this precious substance they'll protect the scale insects from their natural enemies. The way in which the scale insects feed is also a problem because the punctures made by their mouthparts are perfect points of entry for plant pathogens, including bacteria, fungi, and viruses.

The ravages wrought by scale insects are thought to cause considerable economic losses. In 1977 it was estimated the scale insect *Chrysomphalus aonidum* caused damage to citrus crops amounting to $3.75 million in Texas alone. In Florida in 1990, *Unaspis citri* was estimated to cause economic losses to the citrus industry totaling $7 million. In 1975, the economic losses in the Californian citrus industry caused by a number of scale insect pests were estimated to be $22.8 million. In 1982, scale insects were estimated to have been responsible for $37.8 million of losses in Georgia due to damage of ornamental plants, lawns, and turf. In 1990, it was estimated that all the pest scale insects in the United States were responsible for economic losses in the region of $5 billion. If this is even a moderately accurate estimate of the size of the scale insect problem, the global losses due to these insects must be enormous.

Scale insects are serious pests of many crops, but there are many things growers can do to control their numbers and limit the damage they cause. The simplest ways of keeping scale insects in check are cultural techniques, which include ensuring the plant has enough nutrients and water to reduce the impact of scale insect attacks and pruning plants, and/or removing the most heavily infested areas, which opens the plant canopy rendering the scale insects vulnerable to desiccation. In situations where new plants are brought in from other regions or countries it is very important to ensure that all plants have been rigorously checked for any signs of scale insect activity.

Biological control refers to the release of the natural enemies of scale insects. Although this technique is hit-and-miss, it is at its most successful in scale insect control because these pests are more or less sedentary and also because they are often a problem in closed growing situations, such as glasshouses. A variety of parasitic wasps and predatory bugs, beetles, and lacewings have been employed as biological control agents of scale insects with varying degrees of success. The parasitic wasps are the most widely used and successful biocontrol agents of scale insects. Another important consideration in the biological control of scale insects is limiting the degree to which ants can access and protect them, leaving the pests more vulnerable to their predators.

Insecticides can also be used to control scale insects, but this is the most environmentally unsound way of controlling these pests, albeit the most widely used. The insecticides can be simply sprayed all over the infested plants or injected into the plant, so the toxins are taken up by the sap-sucking scale insects. Controlling scale insects with insecticides can be successful, but this is a rather short-term approach as the more the compounds are used, the more the target organisms build up resistance to them. Insecticides also kill nontarget organisms, some of which are predators of the damaging scale insects. Successful control of these important pests relies on a combination of measures where cultural and biological techniques can be combined with the careful and judicious use of chemicals.

FURTHER READING

Ben-Dov, Y., and C. J. Hodgson. *Scale Insects: Their Biology, Natural Enemies and Control,* Vols. 1 and 2. Elsevier, Amsterdam, 1997.

Miller, D. R., and J. A. Davidson. *Armored scale insect pests of trees and shrubs (Hemiptera: Diaspididae).* Cornell University Press, Ithaca, NY, 2005.

Screwworm

There are a number of fly species whose maggots plague a large variety of hosts, making them a serious problem. Perhaps the most important of these are the screwworms, insects of the Old and New World that cause

Screwworms are one of the few insects for which an eradication program has been successful. (Johnny N. Dell)

myiasis (the technical term for fly maggot infestations) in almost any mammal and which have been the target of an ingenious form of control known as the sterile insect technique (see sidebar).

Three species of screwworm are known: the primary screwworm (*Cochliomyia hominivorax*), the lesser screwworm (*C. macellaria*), and the Old World screwworm (*Chrysomyia megacephala*). These species all have a similar life history, but *C. hominivorax* is by far the more important from a human and animal health perspective. The life cycle of these insects and the effect they have on their hosts is very interesting, albeit stomach-churning. The primary screwworm starts off as batches of 200–400 eggs deposited in overlapping layers on a suitable host, which can be something as small as mouse all the way up to very large ungulates, such as cattle. The larvae hatch in 12–21 hours, but unlike the young of warble flies (see warble flies entry), screwworm maggots cannot penetrate the skin of their host. Instead they can gain access through mucus membranes and the smallest of openings, so the female deposits her eggs near the eyes, nose, mouth, genitals, and any wounds or natural openings, such as fly bites, tick bites, the base of the umbilical cord of newborn mammals, scratches caused by vegetation and barbed wire—just about any break in the skin of the host.

Once through the skin, the larvae can feed on the flesh beneath. Because there are many of them they can quickly cause a considerable wound that attracts yet more adult screwworms and opportunistic fly species whose maggots feed alongside the screwworms. Some of the tissue in the growing wound begins to rot and yet more fly species, able to utilize this resource, are attracted. Untreated, heavy infestations are often fatal and even smaller infestations can be deadly because of secondary bacterial infections. In the event that a host survives an infestation of screwworms the propensity of these flies to go for mucus membranes can often result in grotesque deformities of the head and genitals. The screwworm maggots complete their development in five to seven days, at which time they leave the wound, fall to the ground, and burrow into the soil to pupate. Depending on the temperature the adult flies emerge between seven days and two months later to begin the cycle all over again.

In the Americas screwworms feed on all kinds of animals, including livestock, pets, and quite commonly humans, with very unpleasant results.

Sterile Insect Technique

The primary screwworm was such a problem in the United States that developing a means of controlling and even eradicating the fly had the potential to drastically improve livestock production throughout the Americas. In the 1950s, two entomologists, Drs. Raymond Bushland and Edward Knipling, were involved with researching ways of controlling the screwworm. During their studies these two scientists noted that female screwworms only mated once in their life. Therefore, it was reasonable to assume that if the male flies could be manipulated in some way to make them infertile without affecting their behavior, the reproductive cycle of these flies could be cut, ultimately allowing their eradication. Making the male flies infertile was achieved by irradiating them with gamma radiation, a complication of which was working out the dosing and timing of radiation that sterilized the males without compromising their behavior. Perhaps the biggest challenge was rearing the male flies in sufficient quantities so that a release could cover any given area to such an extent that all the wild female screwworms mated only with sterile, irradiated males.

Rearing and releasing sterile screwworms is an ongoing concern. Rearing involves purpose-built facilities where the maggots feed on an artificial diet of blood and the milk protein, casein. When the larvae are ready to pupate they drop to the floor of the rearing chamber and form puparia. As the sex cells in the developing fly are beginning to form (about five days into pupation) the puparia are irradiated, rendering the adults male flies infertile but behaviorally normal. The reared flies are then released over screwworm-infested areas, spelling the end for the resident screwworm population. The technique was first tested in 1951 with a second test conducted on Curaçao off the coast of Venezuela in 1954, where 150,000 sterile male screwworm flies were released every week. Within three months and four generations of the flies, the screwworm was eradicated from this island. In 1958 the technique was employed in the United States and by 1959 screwworms had been eliminated from Florida. It took a further seven years to eradicate the fly from the entire United States—a very short period of time considering the size and scope of the project. Following the eradication of this fly from the United States, the U.S. Department of Agriculture eliminated the screwworm from Mexico and much of Central America, establishing a permanent 300-kilometer-wide barrier zone in the narrow isthmus of Panama. There is a proposal to maintain this zone indefinitely at a cost of approximately $7 million per year. To this day the sterile insect technique keeps the United States and adjacent territories screwworm-free and is also employed in programs aimed at eliminating the parasite from South America and the Caribbean. The technique has also been used to great effect in the control of some other important pests, including fruit flies and melon flies.

The sheer scale of the screwworm eradication program is impressive. During the initial releases in the United States, 14 million sterile flies were reared every week and by 1958 an aircraft hangar in Sebring, Florida, was converted into a gigantic fly factory, capable of rearing 50 million flies per week released with a fleet of 20 aircraft. In terms of applied science, the brainchild of Drs. Bushland and Knipling is without parallel. They devised a technique to control a very damaging pest that uses no environmentally toxic chemicals, has no effect on nontarget organisms, and can be 100 perfect effective in the right situations. This supreme solution to a seemingly insurmountable challenge saw both scientists awarded the World Food Prize in 1992. This inspired breakthrough is summed up in the following quote from a 1970 issue of the *New York Times Magazine,* which read: "Knipling has been credited by some scientists as having come up with the single most original thought in the 20th century."

Further Reading: Wyss, J. H. Screwworm eradication in the Americas. *Annals of the New York Academy of Sciences* 916(2000): 186–93.

The economic losses to agriculture and the costs of controlling and treating myiasis caused by these flies in pets and humans add up to a very significant financial burden. Screwworms were eradicated from the United States in 1966, but prior to this date it has been estimated that this insect cost the U.S. livestock industry alone around $3 billion every year.

Control of this insect prior to 1958 focused on the use of insecticides primarily to kill the maggots, but the expense of continually applying these chemicals, their detrimental environmental effects, and the emergence of insecticide resistance limited the potential of these chemicals in long-term control strategies. An ingenious means of controlling and eradicating screwworms was developed by two entomologists in the 1950s (see sidebar) and today the screwworm-free zone includes all of the United States, Mexico, and some of Central America.

FURTHER READING

Aiello, S. E. *Merck Veterinary Manual.* Wiley, Hoboken, NJ, 2004.

Catts, E. P., and G. R. Mullen. Myiasis *(Muscoidea* and *Oestroidea).* In *Medical and Veterinary Entomology* (G. R. Mullen and L. A. Durden, eds.), pp. 318–49. Academic Press, San Diego, CA, 2009.

Sheep Ked

The sheep ked (*Melophagus ovinus*) belongs to a family of flies known as the hippoboscids, all of which are enigmatic, pretty mean-looking insects superbly adapted to a parasitic way of life. These bizarre and fascinating flies are known by a number of common names, such as keds, louse flies, bat flies, spider flies, flat flies, and so forth. As adults they range in size from 2–12 millimeters and in most species the males are winged and the females are wingless, although both sexes are fully winged in some species, such as the horse louse fly (*Hippobosca equina*). In some species (i.e., the deer ked—*Lipoptena cervi*), the newly emerged adult has fully developed wings, which it uses to good effect to find its victims, but as soon as its feet are firmly on a host the wings break off at their base and the muscles that powered them are broken down to supply the raw materials for strengthening and growing the legs. All hippoboscids have a rather flattened body enabling them to scurry beneath the fur or feathers of their host and their strong legs are tipped with impressive claws and gripping pads to provide good purchase. Around 75 percent of hippoboscids species are parasites of birds and of those that live on mammals many are specialists on bats. The larger mammals parasitized by these flies include many species of ungulate, not excluding several important domesticated animals.

For the purposes of this book we will look at the most important hippoboscid from an animal health perspective—the sheep ked. This hippoboscid is considered to be one of the most important insect pests of sheep and one that has a global distribution apart from lowland areas of the tropics. Its relationship with sheep is probably very old indeed, certainly extending back to the time before the ancestors of these animals (thought to be the mouflon—*Ovis orientalis*) were domesticated in Mesopotamia at least 9,000 years ago. The modern breeds of sheep are excellent hosts for these flies, specifically because they have been selectively bred to produce ever greater quantities of wool, a pelage that affords an ectoparasite an excellent grip as well as being good for hiding.

Like all hippoboscids the sheep ked is remarkable in that it extends maternal care far beyond that of most insects by producing a fully formed larva ready for the rigors of pupation instead of simply laying eggs. Development of the larva takes place in the female's uterus, where it is nourished by its yolk store initially and then by secretions from a pair of milk glands for about 7–8 days in total. Not only does the larva receive all the nourishment it needs to complete its development, but it is also inoculated with the symbiotic bacteria (*Bartonella melophagi*) that will enable it to efficiently process blood meals as an adult. The female sheep ked glues her massive offspring to the wool of her host and here it pupates to produce an adult ked in 19 to 36 days, depending on the season. Female sheep keds are long-lived insects—as you would expect for an insect that only produces one offspring at a time—and in her 4–6-month life span she'll give birth to 10–20 larvae.

Adult sheep keds feed by sucking the host's blood, which they do every 36 hours or so, and this is why this species is of veterinary importance. Small numbers of sheep keds are of negligible importance, but heavy infestations can cause significant economic losses for sheep farmers. Sheep with lots of keds put on less weight and produce less wool than those sheep with no or few keds and there is also the danger of secondary infection of the ked bites. Ked bites can also damage the skin of the sheep, causing scars and small lumps known as cockles. These can often be severe enough to ruin the hide for commercial sale. In the United States alone, hide damage due to sheep ked bites amounts to several million dollars annually. Keds are even thought by some farmers to be the reason for back loss, where adult sheep roll on their back, can't right themselves, and suffocate from the pressure of their internal organs on their diaphragm. Perhaps this behavior happens because the sheep are trying to rid themselves of the parasites or alleviate the irritation they cause. Keds are specialized parasites of sheep, but they are not averse to sucking the blood of humans. Anyone working with sheep will have felt the bite of a ked, which is said to be as painful as a yellow-jacket sting, although the individual response to the bite of this fly varies.

As sheep keds hit farmers where it hurts—the pocket—control of this insect has been thoroughly investigated. Several strategies can help to reduce the population of this parasite in a given flock. One simple method is shearing the flock before the lambing season starts as lots of keds will be killed by the shears and cast off in the fleece, all of which can prevent the flies from moving from mother to lamb. This technique alone can reduce ked populations by as much as 75 percent. When well-timed shearing is combined with insecticide applications (typically pyrethroids) the level of control can be even greater.

FURTHER READING

Lloyd, J. E. Louse flies, keds and related flies (*Hippoboscoidea*). In *Medical and Veterinary Entomology* (G. R. Mullen and L. A. Durden, eds.), pp. 331–45. Academic Press, San Diego, CA, 2009.

Small, R. W. A review of *Melophagus ovinus* (L.), the sheep ked. *Veterinary Parasitology* 130(2005): 141–55.

Tabanids

Tabanids are flies in the family tabanidae and they include the familiar horseflies and the deerflies. The horseflies, especially some species in the genus *Tabanus,* are among the most formidable biting insects—one of the largest North American species, *T. atratus,* a striking, dark-blue metallic fly, is about 30 millimeters long. Both the horseflies and the deerflies have beautifully colored eyes; some species' eyes shimmer with all the colors of the rainbow.

A female horse fly sitting on a horse, preparing to slit the skin and drink the blood. (Paulo De Oliveira /Taxi / Getty Images)

Worldwide, the family tabanidae is represented by around 4,300 species and is at its most diverse in the tropics; however, temperate regions are also home to a large number of tabanid species. There are undoubtedly many tabanids that are still unknown to science, especially in the remote, poorly studied tropical regions. The adults of these flies are difficult to miss and this life stage has been relatively well studied, which is more than can be said for the other stages in the life cycle. We know nothing about the immature stages of many species of tabanid. Because they are almost impossible to rear through successive generations in captivity, the intricacies of their biology, particularly what they do and where they live as larvae as well as their reproductive behavior, pre-egg-laying, are likely to remain a mystery.

Female tabanids deposit their eggs on the ground or more typically on vegetation, normally in tiered clumps. Species in the genus *Goniops* exhibit a remarkable brooding behavior, probably intended to protect their eggs from parasitoids and predators. The female *Goniops* lays hers eggs on the underside of a leaf and she stands over them so her abdomen acts as a roof with her claws penetrating the leaf to provide purchase. Any potential predator approaching the brood is met by the belligerent female, who buzzes loudly and refuses to budge from protecting her eggs. The female *Goniops* finishes her life protecting her eggs and soon after the larvae hatch and fall to the ground, she dies, loses her grip on the leaf, and falls to the ground as well—her maternal duties complete.

The larvae of tabanids live in a large range of moist terrestrial, semi-aquatic and aquatic habitats, including leaf litter, mud, or very moist vegetation at the borders of streams, rivers, and lakes, as well as stream and river beds. Exactly what the larvae do in the wild is very poorly known, but they are fiercely predatory creatures that feed on a wide variety of soft-bodied invertebrates, including others of their kind. Indeed, the larvae are believed to be so rampantly cannibalistic as to be one the most important factors in the natural regulation of their own populations. For pupation, the larvae require slightly drier habitats, so they seek soil that is above the water line. Here some species go through a mysterious ritual of forming a descending, spiral burrow before tunneling up through the center of the spiral to form a pupation chamber. Depending on the species and the environment, the larval development of tabanids can take as long as three years, but pupation is normally completed between four days to three weeks. The adults wait for their bodies and wings to harden in the soil before leaving the earth behind and taking to the wing. Most tabanids are very strong fliers and are able to cover significant distances every day in their pursuit of mates and food. Mating behavior is very poorly known, but once the female's eggs have been fertilized, the race is on to find food so the eggs mature properly.

Like most biting flies, it is only the female tabanids that require blood and this they obtain from a variety of hosts, although large ungulates are the typical victims. Different species feed on different parts of the host; some take blood from the head and others predominantly feed from the legs of their host. All tabanids get their blood meal by slicing the skin and underlying capillaries with their stout, sharp mouthparts. Saliva containing various anticoagulants is introduced into the wound, making it easier for the fly to imbibe the blood. The volume of blood consumed during each meal varies by species, but the largest *Tabanus* species can drink 0.7 milliliters of blood in a single feed and it is not uncommon for tabanids to consume four times their own body weight in blood during each blood meal. The largest tabanids are big insects that buzz noisily in flight;

therefore large herbivorous mammals are only too aware of their presence and they go to great lengths to avoid being bitten, including gadding, keeping away from areas with large tabanid populations, tail swishing, and muscle twitching. The flies are commonly disturbed before they get a chance to penetrate the skin and only a small percentage of tabanid attacks end in the fly getting a bellyful of blood in one attempt. However, these flies are nothing if not tenacious and they keep pestering the same host or other hosts in the vicinity to sate their considerable appetite for blood.

All tabanid species have the potential to be pests because of their need for vertebrate blood, but it is only those species that impinge in some way on livestock rearing or human activities that can genuinely be considered to be a problem. Horseflies and tabanids are considered to be pests because the pain of their bites and their blood-feeding activities are detrimental to livestock rearing as well as being an annoyance to humans. Of greater importance is the ability of these flies to transmit a number of pathogens to humans and livestock.

In areas where livestock are harassed by tabanids on a regular basis the animals are constantly trying to avoid being bitten rather than eating, so weight gain in growing livestock can be reduced by as much as 10 kilograms. In addition to reduced weight gain is the issue of blood loss, which can be 100 milliliters a day in areas where tabanids are common. This doesn't sound like a lot, but sustained over days, weeks, and months it can be very damaging to the overall health of the animal. Most tabanids will quite happily feed on humans as their senses are attuned to seeking out any large vertebrate and our blood is indistinguishable from that of a cow or deer, to one of these flies at least. The bigger tabanids rarely have much luck biting humans, because they are so big and noisy in flight, but the deerflies and smaller horseflies have much more success taking blood from humans to the extent where outdoor activities such as picnics and camping trips have to be curtailed.

Apart from the pain of their bites and the blood they consume, the tabanids serve as vectors for viruses, bacteria, protozoa, and nematodes (see sidebar). For the majority of these pathogens the tabanid is not crucial to their life cycle; therefore, the flies are considered to be mechanical vectors rather than biological vectors. This distinction is little more than academic because the pathogens are no less harmless for the way in which they are transmitted. Tabanids are probably the most effective mechanical vectors of disease because they are large insects and easily interrupted during feeding, meaning they will fly off and seek another host with blood and pathogens on their mouthparts.

The most important diseases caused by tabanid-borne pathogens are loaiasis and animal trypanosomiasis. The former is a disease of humans that is thought to affect 12–13 million people in 11 central and western African countries. The *Loa loa* nematodes are one of the few tabanid-borne pathogens where the fly is actually a biological vector because the nematode requires the insect to complete its life cycle. Immature *Loa loa* nematodes are swallowed by a fly when it takes a blood meal from a human and here they develop until they are almost mature, at which point they migrate to the fly's mouthparts in preparation for infecting a human the next time the fly feeds. In their human host they mature, reaching lengths of 20–70 millimeters (males) and 20–34 millimeters (females), and then breed to complete their life cycle. During their lifetime in the human host, which can be as long as 15 years, adult *Loa loa* wander and it is these migrations that cause the symptoms of loaiasis as they penetrate subcutaneous tissue, eliciting inflammation and swelling wherever they go. Often, *Loa loa* migrate through the conjunctiva and cornea, which can cause considerable discomfort for the victim as well as more serious side effects. In some cases, the nematodes can penetrate deeper tissues, including those of the head, resulting in potentially fatal encephalitis. More information on animal trypanosomiasis can be found in the tsetse entry.

In accord with tabanids' importance in human and animal health and well-being, lots of money and time has been spent in trying to control them; however, they are among the most difficult insects to suppress. There are a number of reasons for this, including the habitats occupied by the larvae (typically subterranean or aquatic in riparian or swampy areas) and the limited contact the flies have with their host (only about four minutes every three or four days), all of which preclude the long-term effectiveness of insecticides. Although conventional control of tabanids using insecticides is very difficult, there are a number of preventative means available to stop the flies from biting in the first place. These include providing livestock with suitable shelters, keeping livestock away from areas where tabanid adults are particularly abundant, especially the transition between woodland and grassland, and flight barriers (tabanid adults prefer to fly around barriers rather than over them, so a two-meter-high enclosure around grazing areas can be effective). Tabanid traps can also be very effective and many designs have been developed and tested, some of which are rather similar to tsetse traps (see tsetse entry). These

Pathogens Transmitted by Tabanids

Pathogen	Disease caused	Vectors
Viruses		
Lentivirus spp.	Equine infectious anemia (swamp fever)	*Tabanus* spp., *Hybomitra* spp., *Chrysops* spp.
Deltaretrovirus spp.	Bovine leukemia	*Tabanus* spp.
Pestivirus spp.	Hog cholera (classical swine fever)	*Tabanus* spp.
Bacteria		
Anaplasma marginale	Anaplasmosis	*Tabanus* spp.
Franciscella tularensis	Tularemia (rabbit fever)	*Chrysops* spp.
Bacillus anthracis	Anthrax	*Tabanus* spp., *Haematopota* spp., *Chrysops* spp.
Protozoa		
Besnoitia besnoitia	Besnoitiosis	*Tabanus* spp., *Atylotus* spp., *Chrysops* spp.
Trypanosoma evansi	Animal trypanosomiasis (nagana)	*Tabanus* spp.
Trypanosoma vivax	Animal trypanosomiasis (surra)	*Tabanus* spp.
Filarial nematodes		
Loa loa	Loaiasis (calabar swelling, fugitive swellings, African eye worm)	*Chrysops* spp.
Elaeophora schneideri	Elaeophorosis (filarial dermatitis)	*Hybomitra* spp., *Tabanus* spp.

traps can be enhanced by baiting them with carbon dioxide or other gases given off by large ungulates (e.g., 1-octen-3-ol and ammonia).

Tabanids have many enemies, especially during their immature stages, and they are attacked and consumed by various fungi, bacteria, protozoa, insects, nematodes, and birds, all of which contribute to the natural control of their populations. Any measures to enhance the populations of these natural enemies are surely helpful in limiting tabanid populations. These measures include habitat management to provide the myriad predatory and parasitoid wasps with nectar sources and places to construct nests, as well as releases of biocontrol agents to supplement the wild populations of natural enemies.

FURTHER READING

McKeever, S., and F. E. French. Fascinating, beautiful blood feeders: Deer flies and horse flies, the Tabanidae. *American Entomologist* 43(1997): 217–26.

Termites

Termites are an amazing and ancient group of insects. Some experts suggest that Permian fossils of insect wings, at least 250 million years old, are clear evidence of just how far back the evolutionary history of these insects extends. Often, termites are called white ants, but this term is completely erroneous. Ants and termites are very different types of insect. Taxonomically, the termites can be considered to be extremely derived, social cockroaches.

Several subterranean carpenter termites on wood. (Michael Pettigrew | Dreams time.com)

Today, around 2,600 species of termite are known to science, but many more species are undoubtedly still to be discovered. Along with the ants, bees, and wasps, they are the only insects to form complex societies based around a queen who is the mother of all the individuals in the colony. These societies behave like a superorganism where different tasks are carried out by specialized individuals. This division of labor gives these social organisms a huge competitive advantage over solitary creatures and as a result they are among the most abundant animals in many parts of the world. In addition to the queen and king, termite colonies are composed of workers, soldiers, and winged alates whose only task is to leave the nest, reproduce, and found colonies of their own. Unique among social insects, the worker termites can be both male and female (in all other social insects the workers are always females).

Termites are included in this book because they have a huge impact on human activities around the globe, particularly in the tropics. However, it must be said that the pivotal role these insects play in the normal functioning of ecosystems vastly outweighs the termite activities that humans perceive to be negative. A relatively small number of termite species are considered to be pests because of the damage they cause to human structures and crops. Broadly, the termites are grouped as dampwood, drywood, subterranean, and arboreal/mound builders. Dampwood termites are restricted in their distribution and as their name suggests they live and feed in very moist wood, such as stumps and fallen trees on the forest floor. Drywood termites are common on most continents and in contrast to the previous group they do not require contact with moisture or soil. Subterranean termites can be very numerous in certain parts of the world and they live and breed in soil, often many meters down. Confusingly, subterranean termites also construct nests above the ground, naturally in trees, but they also use human structures. The mound building species are probably the most well-known termites because of the earthen structures they construct, some of which can be more than eight meters tall!

The first problem associated with termite activity and that which is familiar to most people is their ability to damage human-made structures. Termite species in all the groups mentioned are considered pests and the damage caused by their feeding activities extends to structural timber, other construction materials, household furniture, paper products, many types of synthetic material, and food items. Depending on the species involved, the termites can damage building structures where there is no obvious source of moisture, such as roof beams and wooden paneling, and timber that is constantly exposed to moisture, such as foundation posts in the soil. Typically, signs of damage may not be obvious as the termites often consume the wood from the inside out. By the time the termite infestation is discovered considerable damage may already have been wrought and load-bearing wooden structures may be severely compromised. Where subterranean termites are involved in damage to structures and other materials, their presence is often betrayed by the shelter tubes they construct to protect them from the elements and their predators as they fan out from their underground nests to search for sources of food. These tubes are rapidly constructed from soil, chewed-up wood, and the worker termites' feces. Termites that infest man-made structures, especially the subterranean species, can also interrupt electrical supplies by chewing through insulation, and they can damage pieces of electrical equipment by depositing organic material in and around them. Any such interference with electricity can cause shorting and sparking, increasing the risk of fire.

In the United States termite diversity is quite low and only around 50 species are known; however, this low diversity bears no correlation to the impact of these insects in this country. The economic impact of the structural damage caused by termites is estimated to be at least $1 billion per year and the real figure is likely to be many times higher. At least 90 percent of this damage is attributed to the work of subterranean species in the genera *Reticultermes, Coptotermes,* and *Heterotermes.* Drywood termites in the genera *Incisitermes* and *Cryptotermes* are also known to be structural pests, but they are much less important than the subterranean species. By contrast, South America is home to at least 400 species, a high diversity that complicates the problem as it can sometimes be difficult to identify the termite species in an infestation. The problem in Africa is even more complex as this continent appears to be the center of termite diversity, with at least 1,000 species. Putting a figure to the global cost of termite damage to structures and man-made materials is impossible, but we can be fairly certain it is many billions of dollars every year.

The other major problem that humans have with termites is the damage they cause to food crops and forestry. The tunneling and feeding activities of termites damages seedlings and mature plants and they can also contaminate stored products with fungi. The termite crop pests cause damage in a number of ways, including attacking from the roots, cutting through stems, and tunneling into stems. The net result of all these feeding activities is a reduction in crop yields. The crops damaged by termites vary from continent to continent. In Africa, for example, termites belonging to the genera *Macrotermes, Odontotermes, Pseudacanthotermes, Ancistrotermes,* and *Microtermes* cause damage to groundnuts, maize, sugarcane, yam, cassava, and cotton. These termites can reduce yields of ground nut by 10–30 percent, and maize by up to 60 percent.

Termite crop damage is a nuisance to food producing corporations, but heavy losses can have devastating consequences for subsistence farmers who may have no other way of feeding themselves or their family. Again, it is practically impossible to accurately gauge the economic impact of termite activity on crops, but it must also be many billions of dollars globally every year. Many termite species are specialist wood feeders, so it comes as no surprise that these insects can be serious pests of the agroforestry industry. There are many species of termite that will feed on healthy wood, diseased wood, wood of various ages, and wood from a huge variety of tree species. In the city of Paris alone, one species of termite, *Reticulitermes santonensis,* is known to cause damage worth many millions of Euros every year by munching the many thousands of trees that adorn the streets and parks of this city.

Because of the impact termites have on man-made structures and agriculture there has been a long-term struggle to try and control these insects. A huge number of techniques, varying in expense, practicality, and efficacy, are used to prevent termite damage and to control established infestations. Insecticide applications, especially the potent organochlorine compounds, were a mainstay of termite control for many years, but the devastating ecological impact of these compounds has curtailed their use. In terms of dealing with the termites that cause damage to buildings and other structures, numerous solutions are available. These include designing buildings that are better protected against termite attack, making use of materials that termites cannot use for food and making sure there is at least 30 centimeters' clearance between the bottom of the structure and the soil. Certain types of timber are also resistant to termite damage and nonresistant timbers can be treated with preservatives to keep the termites at bay. Physical barriers are also used to prevent termites from entering a structure and causing damage to the vulnerable materials within. Insecticides are still an important part of termite control and they are applied as soil drenches and baits or dispersed as aerosols to fumigate a closed space. In many respects it is much better for the environment, people, pets, and livestock if termite damage to buildings and material therein is prevented with simple, cost-effective measures rather than being addressed by resorting to the use of potentially damaging poisons.

Several strategies are also regularly used to limit the impact of termites on agriculture and agroforestry. For many decades, the persistent organochlorine compounds were sprayed around cropping areas to act as an anti-termite barrier. This technique is still used today, but as the organochlorines and the related compounds fell out of favor, other insecticides have been used with varying degrees of success. The most effective modern insecticides available are also far beyond the means of most farmers living in developing nations where the termite problem is at its most acute. Farmers do have a variety of cultural techniques at their disposal that are simple and effective, albeit time-consuming. Deep plowing can expose subterranean termites to desiccation and predators and the mounds of certain species can be flooded or burnt to destroy the nest. Another technique simply involves the removal of the termite queen, as the colony soon dwindles and disappears in her absence. Crop rotation, intercropping, and the removal of postharvest debris can also help to reduce the impact of termites in a given location. Plant extracts such as neem (*Azadirachta indica*), wild tobacco, and dried chili have also been used to keep termites away from crops, as has wood ash. Efforts can also be made to render plants more resistant to termites via breeding and ensuring that drought and nutrient stress are kept to a minimum.

Losses from termites in agroforestry can be reduced by selecting low-risk sites in which to plant trees, planting tree species that are suitable for a given region, selecting resistant species or cultivars, reducing mechanical damage that affords the termites entry, maintaining plant health, removing termite nests, increasing biodiversity, and interplanting more than one species. Increasing biodiversity has implications for all pests, as the more diverse a system is, the more chance natural enemies will be present in sufficient numbers to keep the problem species in check.

As our understanding of biology has developed, so has the field of termite control. Biological control and genetic engineering are very active areas of research. Many species of fungi, nematode, and insect have been investigated as biological control agents and to date *Metarhizium* species fungi, *Heterorhabditis* nematodes, and Argentine ants have been shown to be useful in termite control. The potential also exists to engineer the genomes of plants to render them resistant to termite attack. This technology is still in its infancy and there is significant public and expert opposition to the cultivation of transgenic organisms in nature.

Although termites undoubtedly have a huge economic impact around the world we must remember that the fundamental role of these insects in food webs and in ecosystem recycling is essentially priceless. Termites have been around for so long that they are completely integral to the correct functioning of ecosystems throughout the tropics and subtropics. The damage inflicted on our buildings and crops by these insects is a direct consequence of the human population growing in a way that has no regard for the natural world. If the built environment and agriculture were more in tune with the natural world it is highly unlikely that insects such as termites would be much of a problem at all.

FURTHER READING

Abe, T., D. E. Bignell, and M. Higashi. *Termites: Evolution, Sociality, Symbioses, Ecology.* Kluwer, Netherlands, 2000.

Pearce, M. J. *Termites, Biology and Pest Management.* CABI Publishing, Wallingford, United Kingdom, 1997.

Thrips

Commonly referred to as thunderflies, thunderbugs, storm flies, and corn lice, thrips are tiny insects and the species that are known to damage crops are perhaps the smallest of all insect pests. Thrips belong in their very own order—the thysanoptera, which roughly translates as fringe-winged by virtue of their very slender wings, which bear a dense fringe of long hairs. Thrips are so small (usually less than three millimeters) that normal insect wings are surplus to requirements, so they have evolved these delicate structures that bear them aloft. Thrips are believed to be an ancient group of insects: a fossil of an insect that bears many distinctive characteristics of these animals is known from deposits more than 200 million years old.

Worldwide, more than 5,500 species of thrips are known to science, but their diminutive size and secretive ways suggest the actual number must be far higher. Biologically, they are a very interesting group of animals that most people will barely even notice, but as we have seen, the size of an animal has no bearing on how damaging it can be as a pest. Some thrips are winged, while others are wingless. Some species reproduce sexually, while others appear to be parthenogenetic. Although some species

An adult western flower thrip. Some thrips are agricultural pests. They feed by rasping the leaf surfaces and sucking on the plant juices. (Dennis Kunkel Microscopy, Inc. / Visuals Unlimited / Corbis)

of thrips reproduce sexually all of these are known to be haplodiploid, which means that male thrips only have half the number of chromosomes that are found in the females. Because of this chromosomal peculiarity male thrips develop from unfertilized eggs, while females develop from fertilized eggs.

The mouthparts of thrips are adapted for piercing and sucking and most species attack plants, although a number of thrips are predators of other small invertebrates. They are unique in the insect world because they have asymmetrical mouthparts—during their embryological development, the right mandible is resorbed, leaving just the left mandible. Thrips also have a number of resting periods in their life cycle where the immature stages become quiescent and cease feeding before they enter the next instar. These resting stages are sometimes completed in a silken cocoon and for a long time this was assumed to be a period of pupation. Investigations of the thrips' apparent pupa has shown the insect within displays no massive reorganization of its body that so typifies true metamorphosis as seen in the holometabolous insects such as beetles, wasps, flies, and so forth. Another interesting adaptation of thrips is a bladder-like structure (arolium) on each of their feet that can be everted to provide purchase on slippery surfaces.

Thrips may be small animals, but some species are capable of dispersing significant distances, especially if they are carried high into the air, where they may drift for many miles. Most thrips on these long-distance dispersals will die, as such small insects are prone to desiccation, but because many species reproduce parthenogenetically, a population of these insects can be founded by a single individual. It is common for large numbers of thrips to migrate together, particularly when one food source disappears. The senescence and death of flowers late in the season prompts a mass migration of flower-feeding thrips in search of food; hence the name thunderflies, as these plagues often accompany unsettled, stormy weather.

Ecologically, very little is known about thrips. About 40 percent of the known species feed on fungi growing on dead wood and amongst leaf litter—mostly on fungal hyphae but also on fungal spores. A large number of species breed only on grasses, usually in the flowers of these plants, while many species feed only on leaves, some of which induce the development of galls, the abnormal, often distinctive growths that are formed by plant-feeding invertebrates to secure an abundant supply of food and protection from their enemies. Less than 1 percent of the known thrips species are considered to be pests of crops, on which their feeding activities result in plant deformities, scarring, loss of yield, and in some cases, transmission of plant pathogens, notably viruses. It is the species that feed in and around the flowers of their host that cause the most damage because damage to flowers has detrimental effects on the host's fruits. Pest thrips are sap specialists. To obtain this fluid they make an incision in a plant cell with their single mandible and then insert their other mouthparts to pump out the juices of this and adjacent cells.

As with the whiteflies, the damage that thrips feeding causes in commercial agriculture and horticulture is relatively minor compared with the effects of pathogen transmission. One of the most important viruses transmitted by thrips is tomato spotted wilt virus (TSWV), which infects at least 900 species of plant, including many important commercial crops, such as tomato, tobacco, celery, peanut, pepper, bean, potato, and cucumber. At least nine species of thrips are known to be vectors of this plant disease. In India, TSWV is the most important disease of peanuts, where it causes crop losses of between 5 percent and 80 percent. In Hawaii, TSWV can destroy 50–90 percent of lettuce crops during some years. In France and Spain, outbreaks of this disease have caused crop losses in tomato and pepper crops that, on occasion, have been as high as 100 percent.

The control of thrips is difficult. Their small size and proclivity for seeking nooks and crannies on their host plant renders them invisible to all but the most intent observer. Even when thrips are discovered, few people are able to make a definitive identification of these insects. The first sign of a thrips infestation may be the symptoms of the plant diseases they transmit, by which time the population of these pests may be immense. As with all plant pests, prevention is much better than cure, so careful examination of crops can save a lot of time and money by identifying the presence of thrips before a small, isolated population of these insects becomes a problematic outbreak, which is especially true for closed environments, such as greenhouses. Insecticides are the mainstay of thrips control, but as awareness increases of the limitations and drawbacks of these compounds, agriculturalists and horticulturalists are looking for other ways of controlling pest insects. In situations where insecticides are still viewed to be the most effective option, it is important not to use the same active compound repeatedly because doing so risks insecticide resistance emerging among the targeted pests.

Wherever possible it is important to try and harness the natural regulation of thrips numbers offered by natural enemies. Thrips are eaten and parasitized by a huge range of invertebrates and microorganisms. A cultivated environment conducive to the survival of these natural enemies will provide a free and environmentally sound means of controlling thrips populations before they grow to levels where they result in yield losses. In closed growing environments, biological control agents can be introduced to control thrips. Various arthropods have been used in this way, including predatory bugs, parasitic wasps, and predatory mites that between them seek out and consume all the life stages of these insects.

It's worth adding as a final comment that thrips are infinitely more beneficial to humans than they are damaging. Their importance as pollinators alone far outweighs any damage they do to our crops or ornamental plants. Their liking for flowers and their small size means that thrips are very important, albeit overlooked, vehicles for the movement of pollen from flower to flower. The ancient heritage of thrips lends weight to the possibility that they were the first insects to pollinate flowers, thus setting in motion the wheels of evolution that have provided us with the great diversity of flowering plants we know today.

FURTHER READING

Lewis, T. *Thrips as Crop Pests.* CAB International, Wallingford, United Kingdom, 1997.

Terry, I. *Thrips: The Primeval Pollinators?* Seventh International Conference on Thrips Biology. In *Proceedings of Thrips, Plants, Tospoviruses: The Millenial Review* (L. Mound and R. Marullo, eds.), pp. 157–62. Reggio de Calabria, 2002.

Tsetse

Tsetse are among the most infamous of all pests because they transmit the trypanosomes responsible for causing African trypanosomiasis, commonly known as sleeping sickness, as well as a number of animal diseases, such as nagana. These pathogens are closely related to the protozoa that cause Chagas disease in South America and it is not an overstatement to say tsetse and their attendant protozoa have been and remain impediments of development in sub-Saharan Africa.

In the Tswana language of southern Africa, the word *tsetse* simply means "fly." Scientists now recognize 30 species and subspecies of these flies. Most of these are vectors for trypanosomiasis to varying degrees. The most important tsetse vectors of trypanosomiasis are the river-associated species—*Glossina palpalis, G. fuscipes,* and *G. tachinoides*—and the savannah species—G. *morsitans, G. sywnnertoni,* and *G. pallidipes.* These flies have been biting mammals for millions of years and in the past they were much more widespread as fossils of these insects have been found in the 26-million-year-old shales of Florissant, Colorado. Today, tsetse are known only from Africa and isolated populations on the Arabian Peninsula.

Tsetse are quite large as biting flies go—about 7 to 14 millimeters long—and they have an interesting biology, many of the characteristics of which are similar to their relatives, the hippoboscids (see sheep ked entry). A female tsetse mates only once in her lifetime and from this pairing she obtains all the sperm she will need to fertilize her eggs. Like the hippoboscids, tsetse only produce one young at a time and practically all larval development takes place within the confines of the female's uterus. The tsetse larva is nourished by the secretions of a milk gland in the female's

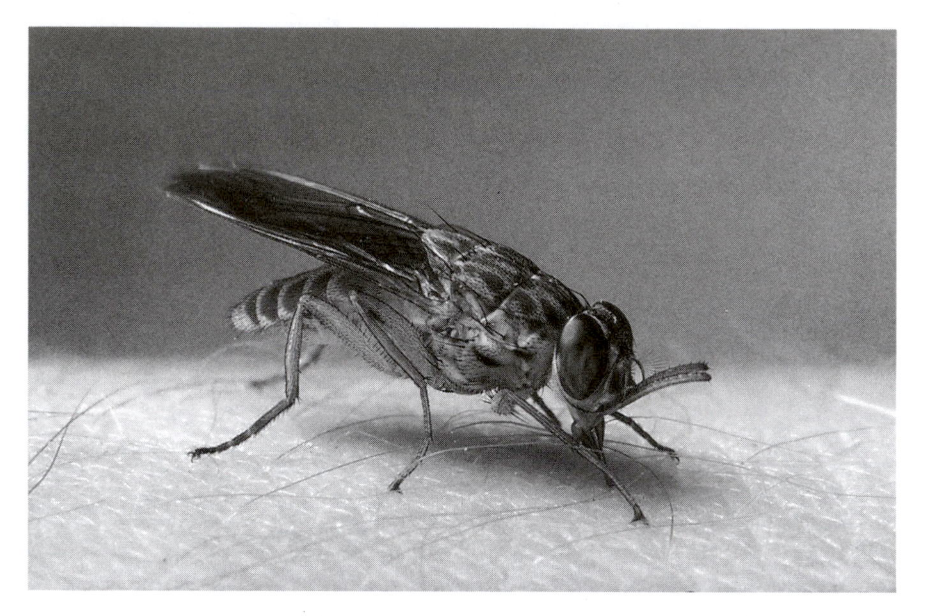

A tsetse on human skin. These flies transmit several pathogens to humans and other mammals. (Oxford Scientific/Photolibrary/Getty Images)

uterus. To breathe, the larva has a number of holes on its rear end, allowing gas exchange with air that permeates the uterus via the outer reaches of the female's genital tract. After about 10 days of this cosseted existence the larva is ready to emerge, so the female finds a suitable spot to deposit her offspring, such as the bare, sandy soil beneath a rocky overhang or tree. Once in a suitable location, the larva edges slowly out, assisted by pushing movements of the female's legs. On the ground, the larva is a sitting duck for all manner of predators, so with rhythmic contractions of its plump body it burrows into the soil. Here its outer skin hardens and darkens to form a puparium that protects the developing pupa. Depending on the species and temperature it takes anywhere between 22 and 60 days for the adult fly to emerge. Female tsetse produce between 8 and 20 larvae in their lifetime, which doesn't seem like many, but each larva is well developed and is ready to pupate when it parts company with its mother. This is in stark contrast to the reproductive strategies of most insects, which produce large numbers of eggs to offset the losses associated with the rigors of development in the outside world.

Unlike many bloodsucking flies, both the male and female tsetse feed on blood. Their typical behavior is to wait on a shaded perch, often on the edge of forests or plantations, for a suitable host to appear. Host preferences of tsetse vary according to the species. Some tsetse species appear to be quite picky in their choice of host and will only feed on a narrow selection of wild animals, such as wild pigs. These species are of minimal importance to human and livestock health. Other species are more catholic in their tastes and it is these that are the greatest danger to humans and domestic animals as they will feed on any suitably sized mammal whenever the opportunity arises.

To locate their prey, tsetse depend on vision and smell. Large, moving objects and the presence of carbon dioxide are particularly attractive to these flies. Interestingly, blue and black, especially royal blue, are very attractive to these insects, the exact reason for which is unknown, but it has been suggested that, to tsetse eyes, blues are the colors of shade. Therefore, when tsetse move toward blue or black objects it is probable they are seeking shade. If their eyes and smell receptors draw them to a host, tsetse are stimulated to feed by the presence of certain chemicals, sensed by receptors on their feet. Once a tsetse alights on a suitable host it uses its sharp mouthparts to pierce the skin and get at the blood in the capillaries. Feeding takes 1–10 minutes, in which time the fly can drink two to three times its own body weight in blood. Massively engorged with blood, the fly must leave the host and seek somewhere safe and shaded to digest its blood meal. In this bloated state the fly's top flight speed is reduced from a brisk 7 meters per second to a lethargic 1.6 meters per second, rendering it very vulnerable to predation, so the priority is to quickly rid itself of the excess weight. To do this the fly has a remarkably efficient excretory system that removes the unwanted water and salts from the plasma in the blood meal, leaving the cells that contain the nutritious fats and proteins. Within 30 minutes after feeding the distended fly has discharged a volume of water and salts equivalent to its own unfed weight.

Tsetse are fascinating insects on many levels, but their proclivities for sucking the blood and transmitting pathogens has brought them into direct conflict with humans. The pathogens of importance transmitted by tsetse are protozoa, specifically trypanosomes. Several species of tsetse-borne trypanosome are known, but the diversity of these organisms in Africa, as well as how they are related to one another, is poorly understood. Of prime public and animal health importance is *Trypanosoma brucei,* which, depending on where it is found and in what host, is identified as *T. brucei gambiense, T. brucei rhodesiense,* and *T. brucei brucei.* These are the causative agents of West African trypanosomiasis, East African trypanosomiasis, and most cases of the animal disease, nagana, respectively. More than 60 million people in 36 sub-Saharan countries—an area of 10 million square kilometers—are directly at risk from African trypanosomiasis and it is estimated that 300,000–500,000 people are infected with the disease. In addition, nagana threatens more than 46 million cattle with an estimated annual cost to the African livestock industry of $1.34 billion.

Like the South American trypanosome that causes Chagas disease, the causative organisms of African trypanosomiasis have a very complex life history. *T. brucei,* after being ingested by a tsetse in a blood meal, develops initially in the gut of the insect before passing to the proboscis and finally maturing in the salivary glands, at which point they are ready to infect another mammal. In mammals, including humans, these parasites multiply in the blood and lymph, resulting in various symptoms, including anemia, edema, fever, paralysis, and eventually death. Without treatment the prognosis is very poor, but drugs are now available that destroy the parasites without severe side-effects.

Although African trypanosomiasis can be treated with drugs, preventative measures are more successful and ultimately cheaper. Insecticides have been used successfully to kill the adult flies, but the evolution of resistance and the potential environmental damage caused by these chemicals are problems that can't be ignored. More inventive and cost-effective is the use of myriad types of trap that attract and catch or kill the flies, thereby stopping them from biting humans and livestock. Many of these are simple in design and easily constructed from local materials. They can be placed in areas frequented by hungry tsetse. In addition to insecticides and sprays, efforts have also been made to control tsetse with the sterile insect technique (see screwworm entry) because female tsetse mate only once, making them ideal candidates for control using this technique. Various complicating factors have prevented the sterile insect technique from achieving anywhere near the same level of success in controlling tsetse in controlling the screwworm.

The tsetse-borne protozoa and the diseases they cause are a massive problem in sub-Saharan Africa in that they impede development. However, they are one of the reasons why much of Africa is free from intensive agriculture and therefore still retains much of its spectacular biodiversity (see introduction). The tsetse problem beautifully exemplifies the complexity of pest science.

FURTHER READING

Gee, J. D. Diuresis in the tsetse fly *Glossina austeni. Journal of Experimental Biology* 63(1975): 381–90.

Krinsky, W. I. Tsetse flies (Glossinidae). In *Medical and Veterinary Entomology* (G. R. Mullen and L. A. Durden, eds.), pp. 289–301. Academic Press, San Diego, CA, 2009.

Steverding, D., and T. Troscianko. On the role of blue shadows in the visual behaviour of tsetse flies. *Proceedings of the Royal Society of London, Series B.* (Suppl.) 271(2004): S16–S17.

Warble Flies

Large grazing mammals are a magnet for flies of every description. Some of these insects come to drink the ungulate's bodily fluids, some gather to make use of the abundant dung produced by these animals, and there are even some flies that use livestock as a nursery for their young. It is to this latter category that warble flies belong. They are large flies bearing a striking resemblance to small bumblebees. Warble flies, also known as gadflies, heel flies, cattle grubs, and ox warbles, are primarily parasites of cattle and Old World deer, although they can also use horses and occasionally humans as hosts. From an animal health perspective, the most important warble fly species are *Hypodermis bovis* and *H. lineatum.* Both of these are natives of Eurasia, but have since been introduced to wherever cattle are reared, making them serious pests in at least 50 countries around the world. These insects are universally feared by the large mammals they depend on, so much so that the distinctive buzzing of these flies around cattle can send the animals into a blind panic where they injure themselves by running into trees, fences, and water. This behavior is known as gadding, hence the name, gadfly. Why these insects elicit such a response is easy to understand when you know a little about their life history.

Adult warble flies have no mouthparts; therefore they cannot feed and the energy reserves they accumulate as larvae only last for about five days, so the race is on for the females to mate and find a host as quickly as possible. To avoid sending potential hosts into a panic the female warble fly approaches very cautiously, often on the ground, in a series of hops until it reaches the cow and crawls up its legs. Carefully, the female lays numerous, 1-millimeter-long, pallid eggs that look like miniature grains of rice.

A close-up of a warble fly. (Shaun L. Winterton)

Each of these eggs is attached to the hairs of the host by a small stalk. Within a week, tiny larvae have hatched from the eggs and make straight for the skin of the beast where they delve into a hair follicle, employing digestive enzymes and their paired mouth hooks to break through the skin to the tissue beneath. There, underneath the tough hide of the animal, they embark on a fascinating and mysterious migration. Using their mouth-hooks they excavate a tunnel in the flesh of the host, growing as they feed on the nutritious muscle and fat. They slowly but steadily make their way towards the head of the animal but when they reach the esophagus they a rest for a while and then make an about turn for no particular reason and head for the rear of their massive host. They tunnel back to the rear of the animal through the muscles of the back. When the larvae arrive at the lumbar region of the host's back they are about 10 millimeters long. They cut a small hole in the animal's hide, through which they thrust the breathing tubes on their hind end. In this position, head down in the flesh of their host, the feeding larva produces a very obvious, raised lump commonly known as a warble. The larvae continue to feed and grow, held in place by a number of spines on their bodies, and when they mature, at a length of around 30 millimeters, they take their leave of the host and fall to the ground. The big wide world is no place for succulent grubs, so they quickly burrow into the soil and undergo metamorphosis in an earthen chamber. Pupation can take two to eight weeks and at the end of it the adult warble flies emerge to seek out more hapless hosts.

From a purely zoological point of view warble flies are fascinating animals, perfectly adapted to exploit large mammals as food, and there is still much we don't know about them. However, to the agricultural industry they are nothing more than a troublesome pest. They are held in contempt by farmers because their presence alone can scare livestock to the extent where less time is spent eating, ultimately reducing the rate at which they put on weight and the amount of milk they produce. The migrating larvae make large cuts of meat worthless as the tunnels fill with what is known as butcher's jelly. Furthermore, the exiting larvae damage the hides of infested animals. A lesser concern, but still important to those people who work with cattle, is the rare situation when these flies inadvertently parasitize humans. In these cases the effects are often gruesome as the larvae will end up in the head or the spinal column, causing the loss of an eye or paralysis of the legs.

In warble fly zones, infestations can be very heavy indeed. For example, in China, 98–100 percent of cattle and yaks can be infested by these parasites and individual animals can have as many as 400 warbles, many of which are a third species of warble fly, *H. sinense,* which appears to be important in this part of the world. In Mongolia, Tibet, and Morocco the intensity of warble fly parasitism can be as high as 700 warbles per host. Putting a figure to the economic losses caused by these flies is very difficult, but even as far back as 1965, the U.S. Department of Agriculture reported these flies were responsible for losses of around $192 million in the cattle industry. More recently, this figure has been estimated to be in the order of $600 million per year. In China, warble fly damage to cattle hides alone was estimated to have cost $15 million in 2003. If the cost of warble fly control is added to these estimates we can see these flies account for significant agricultural losses. With these figures in mind it is no surprise that farmers the world over would like to see these insects eradicated.

Warble flies are controlled with insecticides and by direct removal of the larvae from the lumps on the host's back. Insecticides can be applied to the skin of the host to kill the larvae in their warbles and can also be given systemically to kill all fly larvae in the host's body regardless of their stage of development. Manual removal involves carefully squeezing the larvae from their feeding cavities, making sure not to rupture the grubs as this can cause infections and severe immune reactions. Interestingly, it is young host animals that are most susceptible to the ravages of the warble fly. It appears that older animals build up immunity to the larvae. Efforts aimed at controlling these insects have been very successful in many European countries. The United Kingdom, Ireland, France, Germany, Switzerland, Denmark, the Netherlands, and the Czech Republic are all now free of warble flies.

FURTHER READING

Boulard, C. Durably controlling bovine hypodermosis. *Veterinary Research* 33(2002): 455–64.

Catts, E. P., and G. R. Mullen. Myiasis *(Muscoidea* and *Oestroidea).* In *Medical and Veterinary Entomology* (G. R. Mullen and L. A. Durden, eds.), pp. 318–49. Academic Press, San Diego, CA, 2009.

Whiteflies

As their common name suggests, these tiny insects bear a strong resemblance to flies, but in the insect world appearances can be deceptive. Like the aphids, scale insects, and mealybugs, whiteflies are actually a type of true bug, technically belonging to the family aleyrodidae. Small and delicate, these insects have large wings dusted with a powdery white wax, which they secrete, and the sucking mouthparts that characterize all true bugs. Globally, around 1,500 species of whitefly have been described, but their small size (never more than three millimeters) means there must be many species that are still to be identified, especially in the humid tropics.

Typically, whiteflies are pests of glasshouse crops, such as cucumbers and tomatoes, but they attack a wide range of plants causing considerable economic losses. The life cycle of these insects begins with the female depositing her eggs (as many as 250) in a small circle or crescent on the underside of a leaf. Each of her eggs is attached to the leaf by a small stalk that probably serves to keep the eggs out of the way of predatory insects. Like many plant-feeding hemiptera, the whiteflies have a very interesting reproductive biology. The adult female is able to produce viable unfertilized eggs that develop into males, while the fertilized eggs she produces develop into females.

An oval-shaped nymph, equipped with legs and antennae, hatches from the egg and wanders off in search of suitable feeding sites to probe with its piercing/sucking mouthparts. This first instar nymph sheds its skin and gives rise to a more sedentary second instar nymph with shorter legs and antennae. A third and fourth instar follow. The fourth instar whitefly nymph feeds initially, but then it stops and secretes a case adorned with waxy filaments. This is a resting stage for the whitefly and it is from this case the winged adult eventually emerges. The winged adults disperse, intent on mating, a process that is preceded by some complex courtship behavior. Whiteflies are delicate insects vulnerable to the desiccating effects of the sun and wind. Therefore the whole life cycle is played out primarily on the underside of the host plant's leaves.

As with all plant-feeding hemiptera, whiteflies are considered to be pests because their feeding drains the plant of vital fluids and nutrients. Also, the copious honeydew they produce coats plants, attracts ants, and encourages the growth of fungi. In addition, certain whitefly species transmit more plant viruses than any other hemipteran, plant-feeding pests.

A young adult whitefly on the leaf of a hibiscus. (iStockPhoto)

The impact of these viruses on commercial crops is far in excess of the damage caused simply by the feeding activities of the whiteflies.

One of the more important species is the cotton whitefly (*Bemisia tabaci*), a cosmopolitan species that has been recorded from around 900 host plants, including many important commercial crops, including cotton, sweet potato, cassava, and tomato. Since the early 1980s, the cotton whitefly has been an increasing problem in both field and enclosed agricultural crops and ornamental plants. Heavy infestations of the cotton whitefly and related species (e.g., *B. argentifolii*) can reduce host vigor and growth, cause chlorosis (insufficient production of chlorophyll), uneven ripening, and induce physiological disorders. This species can be a major problem because of the many plant viruses it transmits. To date, the cotton whitefly is known to transmit at least 111 plant viruses belonging to the following genera: *Begomovirus* (Geminiviridae), *Crinivirus* (Closteroviridae), and *Carlavirus* or *Ipomovirus* (Potyviridae). These viruses cause very destructive diseases in commercially important crops around the world, such as cassava mosaic disease (*Begomovirus* spp.), cassava chlorotic stunt virus (*Crinivirus* spp.), carnation latent virus group diseases (*Carlavirus* spp.), and cassava mild mottle disease (*Ipomovirus* spp.). The viruses transmitted by whiteflies can cause crop yield losses of between 20 and 100 percent.

In East Africa, where cassava is an extremely important food crop, the viruses transported by the cotton whitefly, notably cassava mosaic disease (CMD) and cassava mosaic geminiviruses (CMGs), are destroying cassava crops in many countries. In severe outbreaks, cassava mosaic disease can cause root-yield losses of 100 percent and even in the absence of serious outbreaks this plant pathogen can reduce cassava root yield by between 20 and 90 percent in farm fields throughout sub-Saharan Africa. The most important disease of sweet potato in Africa is sweet potato virus disease (SPVD), a complex condition brought about by dual infection with sweet potato feathery mottle virus (SPFMV, transmitted by aphids) and sweet potato chlorotic stunt virus (SPCSV, transmitted by whiteflies). Throughout sub-Saharan Africa, this disease can cause yield losses of at least 98 percent. Sweet potato and cassava are staple foods for many subsistence farmers and any pest or disease that reduces yields can have disastrous consequences for whole communities who live on or near the thresholds of malnutrition.

Another important virus transmitted by the cotton whitefly is tomato leaf-curl virus, which is now an important disease of tomatoes in tropical and subtropical countries. Whenever this virus reaches an area where cotton whitefly is already present, it rapidly spreads through the commercial tomato crop, causing heavy yield losses. In the Middle East, yield losses from this disease have reached 80 percent in some areas. In parts of Australia, entire tomato crops have been lost since it was first reported there in the 1970s. Throughout the Indian subcontinent the cotton whitefly transmits the virus that causes cotton leaf-curl disease, which is capable of causing significant yield losses in this important crop. In the Punjab region of Pakistan alone, cotton leaf-curl virus was estimated to have caused yield losses representing 7.4 million bales of cotton between 1993 and 1998. This was a significant proportion of the region's total cotton production for that period and was worth about $5 billion.

The greenhouse whitefly (*Trialeurodes vaporariorum*) has been a problem for growers of greenhouse crops for many years around the world and its impact as a pest appears to be growing both in greenhouses and also in field crops. This whitefly species can reduce plant productivity and longevity and transmit a number of potentially devastating viruses in a diverse range of crops such as tomato, lettuce, strawberry, cucumber, squash, and pumpkin. A virus transmitted by the greenhouse whitefly—tomato infectious chlorosis crinivirus—can cause severe losses in tomato crops, as experience from Orange County, California, demonstrates. Growers in this region reported losses of $2 million as a result of this disease in 1993 alone, the year when the disease was first identified.

Controlling whiteflies is extremely difficult. Their small size, ability to fly, and propensity for lurking on the underside of leaves means a small infestation on a plant can quite easily escape detection. This capacity to hide is particularly important for commercial greenhouses where whiteflies can be brought in with plant material. In these closed environments where temperature and humidity are kept high and constant, whitefly numbers can explode and cause significant damage. With that said, whitefly populations in greenhouses are also easier to control than populations in field crops.

The most obvious way of preventing or controlling a whitefly problem is careful inspection of plants, especially those brought in from other areas. Insecticides are also commonly used to kill whiteflies, but these insects, like so many other pests, are becoming increasingly resistant to these compounds. Insecticides based on plant preparations can be effective at controlling whiteflies and they also have less detrimental effects on the environment. In greenhouses, sticky traps can also be very effective at controlling whitefly infestations. These are nothing more than bits of yellow card or plastic coated with glue. The adult whiteflies are attracted to yellow and end up getting snagged in the adhesive. One of the most effective ways of controlling these pests in greenhouse environments is with the biological control agents, of which many are now commercially available. A tiny parasitic wasp, *Encarsia formosa,* can be very effective at regulating whitefly populations. Each adult female *E. formosa* can parasitize up to 100 whitefly nymphs. A small ladybird beetle, *Delphastus pusillus,* is also used to control whitefly populations in greenhouses. Both the larvae and adults of this beetle are voracious predators of whitefly nymphs and adults—a single larva of this predator can consume as many as 1,000 whitefly before it is ready to pupate.

Putting a figure to the global economic losses caused by whiteflies and the viruses they transmit is extremely difficult considering the huge variety of crops affected and the cosmopolitan distribution of these insects; however, the loss must be tens of billions of dollars annually. The crop losses directly and indirectly attributed to these insects are nothing more than an economic burden in developed nations. In developing nations, however, where whitefly-borne viruses can devastate staple food crops, yield losses do not just mean lost dollars, but also malnutrition and lost lives.

FURTHER READING

Byrne, D. N., and T. S. Bellows. Whitefly biology. *Annual Review of Entomology* 36(1991): 431–57.

Capinera, J. L. *Encyclopedia of Entomology,* Vol. 4. Springer, Dordrecht, 2008.

Hill, D. S. *The Economic Importance of Insects.* Chapman & Hall, London, 1997.

Pimental, D. *Encyclopedia of Pest Management.* CRC Press, Boca Raton, LA, 2002.

Molluscs

Oyster Drills

Oysters are a delicacy around the world and because of this, aquaculture of these animals is a multibillion-dollar industry. In 2007 alone the global Pacific cupped oyster (*Crassostrea gigas*) aquaculture industry produced 4.2 million tonnes of this species, worth just over three billion dollars. Oysters begin life as free-swimming larvae; tiny, soft-bodied creatures with lots of enemies; however, when they find a suitable place to settle they grow a pair of shell valves that offer protection from most predatory animals, with the exception of the drills—marine snails that are specialist predators of other molluscs, such as oysters. Two drill species are important in the oyster industry: the Atlantic oyster drill (*Urosalpinx cinerea*) and the Asian drill (*Ceratostoma inornatum*). If you're an oyster you don't want one of these snails on your back.

These predatory snails have a special gland on the front part of their muscular foot that exudes an acidic secretion powerful enough to slowly dissolve the calcium carbonate in the oyster's shell. These corrosive secretions in combination with the snail's rasping radula (the snail equivalent of a tongue) enable these predators to bore through their prey's shell to get at the succulent organs and muscles within. Drilling can take some time, but let's face it, the oyster isn't going anywhere. The gland with its secretions is pressed against the shell for 30–40 minutes and the radula is then brought to bear for about a minute to scrape away at the softened shell, a cycle that is repeated for as much as eight hours to penetrate a shell about two millimeters thick. Once the shell has been breached, the snail pokes its flexible proboscis into the hole and proceeds to rend the oyster's body with its scouring radula. For the oyster there can be no escape as it is slowly torn to pieces by the snail.

The Atlantic drill is a native of the Atlantic coast of North America and has been inadvertently introduced to the North American Pacific coast, the southern United Kingdom, and the Netherlands, probably in shipments of oysters. In the United Kingdom the Atlantic drill can kill

An Atlantic oyster drill caught in the action of laying eggs. (Courtnay Janiak)

around 50 percent of the young oysters known as spats, but in its native range mortality of spats due to this species of drill can be 60–70 percent. The Asian drill is mostly a problem in Pacific oyster fisheries. A native of Asia, this drill has also been introduced into other areas and today it can be found along the west coast of the United States and in the oyster beds around the coast of France—the fourth biggest oyster-producing country in the world. In commercial oyster operations this drill can cause spat mortality of around 25 percent, a 20 percent increase in production costs, and as much as a 55 percent drop in profits, all of which amount to a major problem for oyster farmers. Not only do these snails cause considerable losses for the oyster industry, but when they're accidentally introduced to an area they can cause declines in the native mollusc populations by outcompeting them for food and space.

Controlling these marauding molluscs is far from easy, but the most obvious practical measure is reducing their spread by thoroughly checking shipments of oysters for the eggs, young, and adults of Atlantic and Asian drills. Secondly, if these snails are a problem in an area, hand collecting and tile traps can reduce their populations and their impact on the oyster harvest. Thirdly and most controversial is the use of chemicals, unpleasant substances that also go by the name of antifouling agents as one of their uses is to prevent marine organisms from adhering to boats and marine structures. Even at very low concentrations these compounds, such as tributyl-tin, interfere with the hormonal control of sexual characteristics, causing a condition known as *imposex* where female animals develop male sexual organs, leading to reduced fertility and premature death. These chemicals are now banned in many countries.

FURTHER READING

Gibbs, P. E., B. E. Spencer, and P. L. Pascoe. The American oyster drill, *Urosalpinx cinerea* (Gastropoda): Evidence of decline in an imposex-affected population. *Journal of the Marine Biological Association of the United Kingdom* 71(1991): 827–38.

Ruppert, E. E., and R. D. Barnes. *Invertebrate Zoology* (6th ed.). Saunders College Publishing, Fort Worth, TX, 1994.

Slugs and Snails

The mollusca is a hugely diverse animal phylum, including animals as disparate as the mussel and giant squid. However, with all their diversity the vast majority of molluscs are aquatic animals and relatively few species have attempted to conquer terrestrial habitats. The only molluscs with a foothold on the land are the slugs and their very close relatives, the snails.

Snails are distinctive for their helical shells, some of which are very ornate. Zoologically speaking, slugs are essentially snails that have secondarily lost their shell. Indeed there are certain species of slug that retain a vestigial shell on their back or under their skin. Like all molluscs, the body of a slug or snail is very soft and far from waterproof—a prerequisite for most terrestrial animals. They get around the constant danger of desiccation by seeking out moist microhabitats, restricting their activity to times when the air is cool and moist (night), retreating into their shell, aestivating when conditions are least conducive to their survival, and secreting mucus from their entire body. With these behavioral and physiological

The snail's shell helps it to conserve moisture and keep its predators at bay. (Hannu Liivaar | Dreamstime.com)

adaptations, the slugs and snails have become remarkably successful in certain habitats and in many areas they are major crop pests.

Morphologically, the slugs and snails are very different from the vast majority of the animals covered in this book. Locomotion is made possible via a muscular foot through which peristaltic muscular waves are propagated, enabling the animal to glide over almost any surface, its passage lubricated by the secretion of abundant, viscous mucus. An obvious sign of their presence is the sight of distinctive, glistening trails in the morning following the forays of these animals the previous evening. Their primary senses are located on two pairs of protrusible stalks—eye-spots borne on the upper pair and receptors for smell and taste on the lower pair. The sense of sight is far from acute, but they are at the very least able to discern the difference between light and dark. In contrast, their senses of smell and taste are very well developed and it is the former they use to detect their food, often from quite considerable distances.

Once a slug has used its well-honed senses to find its food, it brings to bear its unique mouthparts, an elaborate structure known as the radula.

This structure looks like a conveyor belt of tiny teeth. In the same way we may use a file, the mollusc applies this tooth-covered tongue to its food, rasping morsels into its mouth. To survive on land the slugs and snails have dispensed with the standard means of molluscan gas exchange (gills) and have evolved a lung connected to the outside world through a small hole—the pneumostome—which can easily be seen opening and closing on a large slug. Slug and snail reproduction is also peculiar because the majority of species are hermaphrodites, with individuals possessing both male and female reproductive organs. When these individuals come together, one will act as the male while the other acts as a female, and it is not unusual for any given slug or snail to alternate between behaving as a male or female each time it mates.

Compared to many other important agricultural pests, the available information on the slugs and snails is rather scant even and in many ways they are a neglected backwater of crop pest science. Poorly studied though they may be, slugs and snails pose a significant threat to sustainable agriculture. The importance of slugs and snails as crop pests seems to be increasing, especially in temperate areas where wet summers and mild winters allow these animals to thrive. In suitable habitats several hundred individuals may occupy a single square meter of ground. The reasons for this increasing abundance, especially in temperate areas, are not immediately obvious. In some areas the slugs and snails may simply be filling a niche vacated by insect pests as various control strategies have taken effect. Climate change, especially wetter summers and milder winters in temperate regions, may favor the survival of these animals, but a definite link has yet to be demonstrated. In agricultural crops, slugs and snails are pests because of their appetite for vegetation. The importance of slugs and snails as pests in subtropical and tropical regions appears to be less, but certain species, such as the exceptional giant African snail (*Achatina fulica*), have caused significant crop damage in the areas into which they have been inadvertently introduced. This species and other introduced snails are also a problem as they can out-compete native species, altering the delicate balance of the ecosystem.

Slugs and snails can be very destructive agricultural and garden pests because they will consume almost any part of a plant, destroying seedlings or small plants. In heavy infestations crop losses can range between 20 and 90 percent. They can also eat and contaminate stored foodstuffs, even winding up in the packing process, with understandable consternation for the consumer who finds a dead slug in their frozen peas. In horticulture their feeding activities not only damage plants, but the glistening trails of mucus they leave in their path can reduce the value of ornamental plants. These molluscs are also thought to transmit pathogens from one plant to the next, affording the pathogens easy entry by the feeding damage they inflict.

Suppressing a snail or slug infestation is far from straightforward. Chemical control with poisons such as metaldehyde is the traditional approach. Metaldehyde is usually applied to infested ground in the form of pellets and as slugs and snails move over the treated area they absorb the active compound through their very permeable skin. Inside the body of the mollusc, the toxin destroys the cells responsible for secreting the copious quantities of mucus and the animal is doomed. The toxicity of this compound and other molluscicides is a growing concern because little is known of their long-term impact on the environment. On a small scale, various barriers can be employed to keep slugs and snails at bay, including copper rings or foil placed around the base of plants, a metal that strongly deters these animals. In gardens and small agricultural plots the individual slugs and slugs can be hand-picked and disposed of. Baited traps can also be used to lure the molluscs to their death, but again, these are only effective for small-scale control.

Slugs and snails have a raft of natural enemies, which can be harnessed as biological control agents. These include the larvae and adults of various beetles, voracious predatory insects that patrol the ground for their quarry, dispatching significant numbers of slugs and snails in their lifetime. Vertebrates, including various mammals and birds, are also important predators of slugs and snails. These molluscs are also not without their parasites and of these, various species of nematode have been investigated as biological control agents. To date, at least one species, *Phasmarabditis hermaphrodita,* is available in a commercial preparation that can be applied to the infested ground to infect and skill these pests, particularly slugs.

A number of cultural practices can also be used by farmers who may not have the means to pay for chemical control agents or who find their use abhorrent. These include ensuring the cultivated areas are as free from weeds and organic debris as possible, because these offer refuge to slugs and snails. Occasional reduced tillage of the soil can also be helpful as it enhances the populations of ground-dwelling predators, particularly beetles. Enhancing the cultivated environment to make it less attractive to the slugs and snails, but more attractive to their numerous predators and parasites, is an important part of the integrated management of these pests.

FURTHER READING

Barker, G. M. *Molluscs as Crop Pests.* CABI, Wallingford, United Kingdom, 2002.

Zebra Mussels

Zebra mussels (*Dreissena polymorpha*) are striking little bivalves that have earned themselves an infamous reputation. Natives of Eastern Europe and Russia, these small freshwater molluscs have been inadvertently transported around the world and were established in the United Kingdom by 1824, Sweden by 1920, the Great Lakes by 1989, and California by 2008. Exactly how they have found their way to these countries from their native rivers and lakes is something of a mystery, but international trade and the ballast tanks of ocean ships may be to blame as the water to fill these tanks is taken from whatever river, lake, or inland sea in which the ship happens to be docked (complete with any aquatic organisms). Thousands of miles later the water and the accidental passengers are discharged at journey's end. Mussel larvae are tiny, free-swimming animals that swim away from their mother in the hope of finding a good place to settle and grow. It's probably at this stage in their life that they are sucked into a ballast tank and transported hundreds or thousands of kilometers. As the adults can stick to just about any surface with extremely strong adhesive, miniature guy ropes known as byssus threads, it is also possible they may be transported on anchors and other equipment, as they can survive being out of the water for around five days. They may have also dispersed naturally along the many canals and waterways that were constructed during the industrial revolution in Europe.

These animals feed by filtering water and trapping edible particles on soft, mucus-covered structures safely concealed out of sight within their shell. Completely sedentary as adults, zebra mussels waste precious little energy on movement, enabling them to grow rapidly. They lay down layer upon layer of calcium carbonate to form the two parts (valves) of their shell, the only protection they have from predators. Wherever there is suitable substrate for them to fix, the zebra mussels can be present in huge numbers. There are reports of 700,000 zebra mussels being attached to one square meter of substrate. In the areas where they've been introduced the lack of natural enemies allows the zebra mussel populations to reach these unnaturally high levels. It's these massively dense aggregations that make the zebra mussel such a problem as they'll amass on any suitable

Zebra mussels attached to a common clam. (Randy Westbrooks/USGS)

surface, whether it's the water intake pipe of a power station, the underside of a boat or dock, and so forth. Their presence on beaches can be a problem as their shells are sharp enough to cut the feet of bathers. As soon as they've been introduced into an area there is a never-ending battle to remove the adult mussels from important structures.

Not only do they cause problems for industries and recreational activities associated with the waterways, but they also upset the delicate balance of the ecosystems into which they are introduced. Dense aggregations of zebra mussel can smother and out-compete native bivalves for food and render large areas of lake and river bed unsuitable for other aquatic animals. The mussels have even been blamed for causing outbreaks of avian botulism, which has killed thousands of birds. One theory to support these claims is that the zebra mussels accumulate high concentrations of organic pollutants in their bodies because of their filter-feeding. These pollutants find their way up the food chain, becoming increasingly concentrated as they go, until the top predators, usually birds, receive a toxic dose and die.

In contrast, some scientists think these molluscs have some positive effects on the ecosystems where they are now found. The size of their populations and the sheer quantity of water that billions of these animals can filter acts to clarify the water and the feces and other waste they produce is readily accessible to bottom-feeding animals, including several species of fish, the populations of which are increasing in areas with large populations of zebra mussel. This may be perceived as a positive by some people, but one can't ignore the fact that the presence of huge numbers of an alien species has huge repercussions on an ecosystem, many of which go unnoticed until restoring the balance becomes impossible.

Ever since the zebra mussel was first detected in the United States, federal and state agencies, companies, and individuals have been in a struggle to control its numbers and limit its spread. Both of these objectives are proving practically impossible to achieve. Something as simple as moving a small boat from an infested water body to a pristine water body expands the range of the zebra mussel. Even a little bit of water weed harboring some zebra mussel larvae can be transported to a pristine water body on the foot of a duck or goose. There are so many ways for the mussel to spread that seeking to stop it seems futile. Huge amounts of money have been thrown at this problem and it has been estimated that more than $500 million is spent every year managing mussels at power plants, water systems, and industrial complexes, as well as on boats and docks in the Great Lakes. Apart from the tedious task of stripping adult zebra mussels from important structures, chemicals have been poured into the water to kill the adults and the larvae. Surfaces the mussels aggregate on have been painted with substances to prevent them from attaching. Both of these techniques are environmentally unsound as the chemicals in question are very nasty and inflict their own damage on native flora and fauna. It has been found that the mussels don't like attaching to copper-nickel alloys, but covering suitable substrates in this material would only be feasible for small areas and critical structures. Biological control with enemies from the zebra mussel's native range has been touted as a possible means of limiting its spread, but this is fraught with its own difficulties, including not knowing how other alien species will behave when introduced to a new ecosystem. With all this in mind it seems the zebra mussel is here to stay.

FURTHER READING

Ludyanskiy, M. L., D. McDonald, and D. MacNeill. Impact of the zebra mussel, a bivalve invader. *Bioscience* 43(1992): 533–44.

Nalepa, T. F., and D. W. Schloesser. *Zebra Mussels: Biology, Impact, and Control.* Lewis Press, Boca Raton, LA 1993.

Nematodes

Nematode Pests of Animals (Including Humans)

Nematodes are probably the most abundant multicellular organisms on earth, but their small size coupled with the fact they are often very difficult to identify conspire to make them poorly known beyond a few specialist scientists. They are worm-like animals ranging from 0.1 millimeters to 9 meters long, although most species are near the lower boundary of this range. To the casual observer, different species of nematode may look very similar, but their superficial simplicity belies their internal complexity and the huge variation in life histories found in this phylum of animals. Currently, around 80,000 nematode species are known, but they are so poorly known it has been estimated there could be as many as one million species. Some nematode experts have even suggested there could be as many as 100 million species of nematode. Not only are they very speciose, but nematodes are also the most ubiquitous multicellular organisms, being found in every imaginable habitat, from the deep oceans to the soils of arid areas as well as the bodies of other animals. They abound wherever they occur. For example, a single rotting apple can support as many as 90,000 nematodes. A six- to seven-millimeter sample of mud can be home to 1,074 nematodes representing 36 species. Three to nine billion of these animals can be found in a single acre of good-quality farmland in the Unites States.

The diversity and abundance of nematodes in every ecosystem indicates that their importance is probably vastly greater than what we currently appreciate. Although the majority of species have nothing but unseen, positive impacts on our lives, there are many species that affect the health of humans, livestock, and our crops.

The impact of these small, superficially simple animals on human health as well as the menagerie of animals we have domesticated and other animals we depend on for food, such as fish, is considerable. To a parasitic nematode, the body of an animal is a cushy ecosystem with lots of niches

These giant round worms (*Ascaris lumbricoides*) were living in the intestine of a Kenyan child. (CDC/Henry Bishop)

in which to carve out a living. Like their free-living relatives, the parasitic nematodes survive and indeed thrive in a huge range of internal habitats. Many species spend some of their time as juveniles in the blood while the adults populate the gut. There are also many parasitic nematodes that penetrate the muscles and organs to form small capsules in which they find safety and nourishment. There are even those species that wander freely about in the body cavity as adults. There really is an amazing array of ways in which parasitic nematodes exploit their hosts.

Many thousands of nematode species are thought to parasitize vertebrates. The actual diversity of nematodes that take advantage of mammals, birds, reptiles, amphibians, and fish may never be known. These typically small, worm-like organisms are more than likely a consistent feature of vertebrate life, with a heritage that must extend back many hundreds of millions of years. Some of these nematodes are obligate parasites, meaning they can survive nowhere else apart from the body of their host, while others are only parasitic during certain stages of their life cycle. Still more are free-living, but can occasionally be accidental parasites if they find their way into another animal. This latter relationship is probably the way in which nematodes came to parasitize animals in the first place. It's not difficult to imagine an ancestral vertebrate sifting through the sand and silt at the bottom of a shallow sea, searching for its prey, only to accidentally ingest a few thousand nematodes representing a variety of species, some of which went on to form lasting relationships, albeit harmful ones, with vertebrates.

Nematode parasites of vertebrates are so diverse in their characteristics and habits that it would take a small library to cover them in sufficient detail. With this in mind we'll look at some of the more important nematode parasites of humans and domesticated animals.

In the nematode order trichurida, there are two genera of nematodes that are very important in human and animal health: *Trichuris* and *Trichinella*. Nematodes in the genus *Trichuris* are important parasites of mammals (including humans) and they are commonly referred to as whipworms because of their long and slender appearance. They are gut parasites that feed from the intestinal mucus membrane with the front part of their body embedded in the tissue. Heavy infections (more than 100 whipworms) can cause dysentery, anemia, rectal prolapse, physical and mental developmental problems, and even death in rare cases. It is estimated that more than one billion people around the world are infected with this nematode and in some areas of the world, notably east Asia, the prevalence may be as high as 95 percent. An adult female whipworm can produce 3,000–20,000 eggs per day, which exit the body in the host's feces. It is food and water contaminated with egg-laden feces that maintains the cycle of whipworm infections.

Nematodes in the genus *Trichinella* are among the smallest parasitic nematodes, but their diminutive size has no bearing on their ability to cause disease in humans and animals. The route of transmission for *Trichinella* nematodes is the consumption of meat harboring the infective juveniles. The juvenile worms spend some of their time in the mucus membrane of the intestine and here they grow and develop, eventually reaching adulthood. In the intestine the nematodes mate and the females go on to produce hundreds or even thousands of eggs over a 4–16-week period. The eggs hatch and the juveniles find their way into the host's circulatory system to be carried throughout the body to every conceivable type of tissue. Their intended destination is the skeletal muscle, where they penetrate an individual muscle fiber and hijack the cellular machinery within. They manipulate the cell to convert it into a collagen-encapsulated nurse cell complete with an excellent blood supply. Protected and nourished in this tiny host-derived capsule they wait to find their way into another host when the muscle in which they are embedded is eaten. Secluded in their tiny capsules, the juvenile nematodes go into developmental arrest, but remain alive and potentially infective for at least 30 years.

The activity of adult *Trichinella,* and the dispersal and nurse cell formation of thousands of juveniles throughout the body, can cause a number of symptoms in the host, some of which can be very serious. The female nematodes penetrating the intestinal mucosa can cause nausea, vomiting, sweating, and diarrhea. These symptoms are mild compared to those caused by the wandering juveniles, which can cause pneumonia, pleurisy, encephalitis, meningitis, nephritis, deafness, peritonitis, brain or eye damage, and potentially fatal damage to the heart. Formation of nurse cells in the skeletal muscle is associated with extreme muscular pain, difficulty breathing, swallowing or chewing, and heart damage. Fortunately, this parasite is nowhere near as common as it once was. It is estimated that at least 10 million people around the world are infected with this nematode, most of whom are in developing nations. In Western countries cases of *Trichinella* infection are now very rare. Humans are typically infected with this nematode when they consume raw or undercooked meat, especially pork, or if food has been prepared by someone handling infected meat. The potential of *Trichinella* infection is the single biggest reason to avoid the growing trend for consuming raw and partially cooked meat.

Nematodes in the genus *Strongyloides* can be common parasites of humans in tropical and subtropical regions. They cause symptoms by penetrating the skin as infective juveniles, damaging lung tissue during their migration from the point of entry to the digestive tract, and by invading the intestinal tissues. Infection by this species can arise from contact with contaminated food, water, or soil. In some parts of Africa, as many as 48 percent of the population can be infected. In the southeastern states of the United States, prevalence ranges from 0.4–4 percent of the population. Improved sanitation and hygiene and a greater understanding of how such parasites are transmitted means this nematode is now rare in developed nations.

Hookworms are another group of nematodes that are very important parasites of humans and domesticated animals. Two genera, *Ancylostoma* and *Necator,* cause disease by migrating, penetrating tissues, and feeding on blood and tissue fluids from the intestines of their hosts. The juveniles of these nematodes gain entry to their host via the oral route or by burrowing through the skin. If it's the latter they must embark on a migration that sees them breaking out of the circulatory system in the lungs and heading for the digestive tract. Penetration of the juveniles through the skin can caused localized symptoms and juveniles breaking out in the lungs can cause blood loss, which can be serious in massive infections. However, it is the feeding activity of these worms in the intestine that causes the most symptoms. In heavy infections, victims may lose 200 ml of blood every day. Over many months this can have dire consequences for rapidly growing children, especially those who have poor diets lacking in essential nutrients. Hookworm infection is a major public health problem. At least 750 million people around the world are infected with this parasitic nematode and many of these are extremely poor people in developing nations. In some areas, especially sub-Saharan Africa, 80–100 percent of the population can be infected with hookworms.

As their name suggests, the giant roundworms are among the largest nematodes. There are two very closely related species, *Ascaris lumbricoides,* which infects humans, and *A. suum,* which is a parasite of pigs. It is possible that *A. lumbricoides* evolved from *A. suum* following the domestication of pigs by humans, an event that occurred at least 10,000 years ago. *A. lumbricoides* is perhaps the most common animal parasite of humans with at least 1.4 billion people being home to this nematode. In some areas, such as parts of Indonesia, 90 percent of the population may be infected. The adults of this species are denizens of the intestines and the adult females are prodigious breeders, capable of producing at least 200,000 eggs every day, which are expelled with the feces. These eggs are famously tough and can remain viable for at least 10 years in the soil. Contaminated food and water and poor hygiene provide the route of entry for this nematode. Once swallowed the eggs hatch and the larvae commence a hazardous migration through bloodstream and lungs and back to where they started in the intestine. This circuitous route may seem odd, but is probably a behavioral vestige that was crucial to the survival of their ancestors in some long-extinct or bypassed intermediated host. Small numbers of these nematodes in the intestines do little if any damage. Problems arise in heavy infestations because the nematodes can block important channels in the alimentary canal and elsewhere, especially when they become restless and move throughout the body. During these wanderings they can cause considerable mechanical damage in such incongruous locations as the middle ear and their sheer numbers can rupture or block the intestine, with fatal consequences. It is estimated that global annual deaths due to *Ascaris* infections may be as high as 100,000.

Toxocara cani is a nematode parasite of dogs, but it can also infect humans, often with grisly results. The prevalence of this nematode in the dog population can be very high, especially in puppies in the United States, where infection rates may be as high as 100 percent. If dog feces contaminated with eggs find their way into a human, the juveniles, like those of *Ascaris* species, penetrate the intestinal epithelium and get transported through the body in the blood, hoping to reach the lungs so they can complete their development in the intestines. Migrating larvae are what cause problems in humans because they can end up in many organs, destroying lung, liver, kidney, muscle, and nervous tissue. If a juvenile *T. canis* finds itself in the eye, damage and blindness can result.

Pinworms (*Enterobius* spp.) are also extremely common parasitic nematodes. As adults, they are found in the intestine of their host. Females deposit 4,000–16,000 eggs at or near the anus, commonly exiting their host and crawling about on the perianal skin of their host. The damage caused by the adult worms feeding in the intestine and the irritation they cause when they are laying their eggs are the biggest problems associated with these parasites. They are also known to enter the reproductive tract of female hosts where they become encapsulated in granulomas. Worldwide, pinworms probably infect at least 400 million people and in the United States alone there are thought to be 20–40 million people who harbor this parasitic nematode, most of whom are children.

The penultimate group of parasitic nematodes to be covered here are the filarial nematodes that between them infect millions of people around the world. These species have already been covered to some extent in other entries, namely the mosquitoes and the blackflies. These nematodes all use arthropods as intermediate hosts and it is via these intermediaries that they gain access to their definitive, vertebrate hosts. For more information on these nematodes, see the entries on mosquitoes and blackflies.

Finally, we will have a brief look at the Guinea worm. This is an unusual parasite that is probably the only nematode parasite that we have a reasonable chance of eradicating. The Guinea worm uses small freshwater crustaceans as its intermediate host and they gain access to their definitive host—vertebrates—by being inadvertently swallowed. The fully grown females cause debilitation and wounds that are prone to infection. Breaking the cycle of infection by even crudely filtering drinking water provides a simple means of eradicating this parasite. The World Health Organization implemented a global eradication program and to date it has been very successful. In 1989, there were 892,055 cases of Guinea worm. By 2008, the number of global cases had fallen to 4,619.

The representative nematode parasites above illustrate just how important these small invertebrates are in human and animal health. The global economic burden of these invertebrates must be enormous—probably tens of billions of dollars every year in sickness, lost productivity, and preventative and remedial medicine. The potentially serious consequences of nematode infection have attracted a lot of interest in developing ways to control these parasites. The most simple and arguably the most effective way of controlling parasitic nematode populations is by breaking the cycle of infection. In almost every nematode species that parasitizes humans and domesticated animals, the eggs laid by the adult females leave the host. These eggs or the juvenile larvae must find their way back into the host via the mouth or by penetrating the skin. Basic levels of hygiene and good sanitation are two very simple and effective ways of ensuring the eggs of the parasitic nematodes do not find their way into their host. Ridding the body of juvenile and adult nematodes can be achieved by using a number of different compounds, collectively known as antihelminthics (see introduction). Examples of these compounds include avermectins, piperazines, and mebendazole. These can kill the nematodes outright or paralyze the adults so they lose their grip on the intestinal mucosa and are passed out of the host's anus. As a rule, it can be harder to rid the body of the wandering juveniles than it is to eliminate the adults, but this varies depending on the nematode species involved.

FURTHER READING

Anderson, R. C. *Nematode Parasites of Vertebrates: Their Development and Transmission.* CABI, Wallingford, United Kingdom, 2000.

Lee, D. L. *The Biology of Nematodes.* Taylor & Francis, London, 2002.

Roberts, L. S., and J. Janovy, Jr. *Foundations of Parasitology.* McGraw-Hill Higher Education, New York, 2008.

Nematode Pests of Plants

Nematodes are extremely important pests of agriculture, horticulture, and forestry—perhaps *the* most important crop pests of all. Of the various nematode species that parasitize plants, most attack the plant roots and a small number affect the leaves and foliage. Plant pest nematodes rarely cause the death of the plants they attack, but they can reduce crop yields to such an extent that it becomes uneconomical to grow crops where they occur. Every major cultivated crop on the planet is affected in some way or another by nematodes and the small size of these animals and their proclivity for damaging the underground parts of plants means their economic impact is often overlooked. It is estimated that some 10 percent of global crop production is lost because of nematode damage, which represents a colossal economic burden for farmers everywhere.

Plant parasite nematodes are broadly divided into four groups: ectoparasites, migratory endoparasitoids, sedentary endoparasitoids, and semi-endoparasites. The ectoparasitic nematodes remain on the surface of the plant tissues and they penetrate the plant cells with a structure known as a stylet to get at the nutritious contents. These ectoparasitic nematodes can be foliar or root specialists and the latter can have short or long stylets to penetrate shallow or deeper tissues. Migratory endoparasitic nematodes can penetrate plant tissues in all stages of their life cycle, moving through and feeding on the plant tissues. Many of the species in this group move between the soil and the plant roots, and like the previous group there are

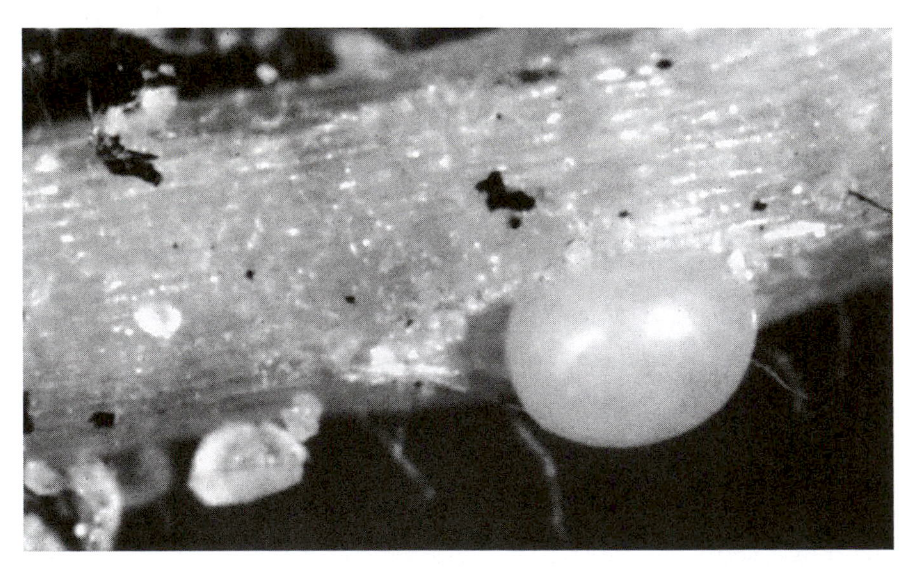

The characteristic golden "cyst" of the potato cyst nematode, a very important pest of this valuable crop. (USDA)

specialists of the roots and foliage. The sedentary endoparasitic nematodes enter their host plant and locate a permanent feeding site where they become immobile and often hugely distended. The sedentary nature of these species can trigger a response in the host-plant tissues that results in the formation of a gall around the nematode. The semi-endoparasitic nematodes partially penetrate their host plant during their immature stages, but the back end of the animal protrudes into the soil. Like the previous group, these nematodes form a permanent feeding site and the portion of their body outside the plant gradually swells.

Nematodes can be a problem for crop growers because their populations can grow rapidly, often enhanced by the ability of some species to reproduce asexually as well as sexually. Nematode feeding reduces plant vigor, especially if their attacks coincide with drought or nutrient stress. Also, the damage they do to the epidermis of plants permits the entry of pathogens that may cause further yield losses and even plant death. Plant nematodes can be very difficult to control and eradicate once they have become established largely due to the very hardy nature of certain stages in the life cycle, which are remarkably resistant to extremes of temperature and humidity and even noxious chemicals designed to kill them. The small size of nematodes coupled with their ability to survive extremes for extended periods of time means they can be easily and invisibly transported from one site to another in the movement of water, soil, and plant material. This is probably the primary route via which nematodes infect regions and continents.

As has already been mentioned, nematodes are ubiquitous pests of crops around the world, but some are more damaging than others. Species such as the potato cyst nematodes (*Globodera* spp.) and the root knot nematodes (*Meloidogyne* spp.) are serious pests of modern agriculture. As their name suggests, potato cyst nematodes are pests of the potato and they originated in the Andes, the home of this plant. Today they are found in many countries throughout Europe, North America, and South America. The economic impact of these species is impossible to accurately gauge, but in the European Union alone the cost of crop losses and antinematode measures are estimated to be many hundreds of millions of Euros every year. Worldwide production of potatoes as of 2007 was 325 million tonnes and the potato cyst nematodes are thought to cause yield losses of around 12 percent. This means that more than 30 million tonnes of potato are lost every year to these miniscule animals. The two potato cyst nematode species (*G. rostochiensis* and *G. pallida*) are considered to

Some of the More Important Nematode Pests of Cultivated Crops, Their Hosts, the Crop Losses They Cause, and Where They are Found

Species	Hosts	Crop losses	Distribution
Potato cyst nematode (*Globodera rostochiensis* and *Globodera pallida*)	Potatoes and other *Solanum* spp., such as tomatoes	~10–12%, but losses of up to 60% have been reported in heavy infestations	Essentially worldwide today
Rice white-tip nematode (*Aphelenchoides besseyi*)	Rice and also strawberries	Up to 50% in some circumstances	Major rice growing areas: Africa, Asia, Eastern Europe, North, Central, and South America, and the Pacific region
Bulb and stem nematode (*Ditylenchus dipsaci*)	Many food plants affected: fava beans, garlic, leeks, lucerne, maize, oats, onions, peas, potatoes, rye, strawberries, sugar beet, tobacco, It celery, lentils, and wheat. Many nonfood plants also affected.	In heavy infestations, crop losses of 60–80% are not unusual	Most temperate regions
Burrowing nematode (*Radopholus similis*)	Many plants attacked, but very important in banana, pepper, and *Citrus* spp. crops	Up to 50% in banana crops and 70–80% in *Citrus* spp. crops	Fiji Islands, Australia, Florida, Central and South America, several Caribbean islands, tropical Africa, and some European countries
Potato tuber nematode (*Ditylenchus destructor*)	Mainly potatoes, but also carrots, ground nuts, garlic, sweet potato, and nonfood crops	Yield losses of up to 40% have been recorded in potato crops	Essentially worldwide
Soybean cyst nematode (*Heterodera glycines*)	Soybean is the most important host, but also attacks many other plant species	In Japan, soybean yield loss is 10–75%	Essentially worldwide

Spring crimp nematode (*Aphelenchoides fragariae*)	Strawberries	20–30% is typical, but can be much higher in heavy infestations	Most temperate regions where strawberries are grown
Pine wilt nematode (*Bursaphelenchus xylophilus*)	Various trees in the genus *Pinus*	30–90%	North America, Europe, and East Asia
Californian dagger nematode (*Xiphinema index*)	Grapevine	20% in some situations	Worldwide
False root-knot nematode (*Nacobbus aberrans*)	Potatoes are the most important host, but cabbage, *Capsicum*, carrots, cucumbers, lettuces, prickly pear, sugarbeet, and tomatoes are also affected.	55–90% have been reported in the South American potato crops	Essentially worldwide
American dagger nematode (*Xiphinema americanum*); probably a group of very closely related species	Many plants, including food crops, trees, and horticultural crops	Difficult to quantify as identification is problematic; probably causes very significant losses to many crops throughout the world	Worldwide
Ear cockle nematode (*Anguina tritici*)	Wheat, rye, spelt, emmer, and rarely barley	30–70%	North Africa, Eastern Europe, and Asia
Sugar beet cyst nematode (*Heterodera schachtii*)	Sugar beet	10–30%	Temperate regions, including North America and 39 sugar beet growing regions
Red-ring nematode (*Bursaphelenchus cocophilus*)	Palm trees	20–80%	Central America, South America, and many Caribbean islands

be such an agricultural threat that the movement of material possibly infected with these animals is heavily restricted in many countries around the world. A characteristic of potato cyst nematode biology that makes them particularly successful pests is the formation of cysts. These tiny, bead-like structures are actually the swollen body of the dead female and each one contains 200–600 eggs. The nematode larvae safe within their eggs inside the durable capsule of their withered mother can bide their time for more than 10 years in a state of suspended animation until a patch of ground is planted with their host plant.

The second example of a serious nematode pest of agriculture is the root-knot nematode, which is not fussy when it comes to host plants, enabling them to damage a wide range of agricultural and horticultural crops. One of the root-knot nematode species (*M. naasi*) attacks a number of cereal crops and is known to cause yield losses of 75 percent in barley grown in California. Another species, *M. artiellia,* has been shown to cause yield losses of 90 percent in wheat grown in Italy. These potato cyst nematodes and the root-knot nematodes clearly exemplify just how damaging these small invertebrates can be.

Crop damage caused by nematodes is a huge, underestimated problem. Because they affect so many types of plant in so many areas, a number of techniques have been devised to try and control or even eradicate them. The most successful means of controlling nematodes is to prevent them from becoming established in the first place. This means imposing strict regulations on the movement of soil, water, and plant material that may harbor adult nematodes, cysts, and eggs. Preventing nematode infestations can also be achieved by certification schemes that provide assurances that seeds and seedlings are nematode-free. In situations where the movement of soil or plant material is unavoidable, a quarantine period will show if symptoms of nematode attack develop. If nematodes do become established in a given area there are several things a farmer can do. Simple, environmentally sound techniques include the removal and destruction of diseased plants. Often, however, farmers and growers will resort to the use of pesticides, some of which are purported to be nematode specific, hence the name nematicides. Nemagon (dibromochloropropane) was a commonly used nematicide applied as a soil fumigant, but it was found to cause sterility in male workers who handled the product as well as persisting for a long time in the environment. The use of this product has since been banned and it is an example of just how damaging some synthetic pesticides can be. More environmentally sound nematode control methods include soil treatment techniques, such as hot water dousing and percolating superheated steam through the ground, both of which kill the various stages of the pest nematodes. Cultural practices such as crop rotation, fallow periods, cover crops, and green manure can also reduce the burden of nematode pests by preventing the buildup of their populations to levels where the damage they cause surpasses economic thresholds.

Technological advances are also allowing the development of high-tech ways of controlling nematode populations. Genetic engineering has the potential to render plants resistant to pest nematode attack. However, splicing the DNA of different species together to produce characteristics that are desirable to agriculturists faces considerable opposition from experts and the public alike. There is also increasing interest in controlling nematodes with biological control agents as they are vulnerable to a range of pathogenic bacteria and fungi. Most interestingly of all are some of the multicellular fungi, which have fascinating means of preying on these sinuous animals. Fungi in the genus *Arthrobotrys* are specialist predators of nematodes and are essentially living lassos. The thread-like hyphae of these fungi are adorned with small constricting rings, through which nematodes occasionally try and squirm. This is the last thing they do, as the ring tightens around the nematode, locking it in a fatal embrace. Eventually the victim dies and the fungal hyphae penetrate the body of the nematode to feed on the tissues within.

FURTHER READING

Bridge, J., and J. L. Starr. *Plant Nematodes of Agricultural Importance: A Colour Handbook.* Manson Publishing, London, 2007.

Lee, D. L. *The Biology of Nematodes.* Taylor & Francis, London, 2002.

Platyhelminthes

Cestodes

Collectively known as tapeworms, cestodes are among the largest and most well known of all the parasites that infect humans and domesticated animals. Humans have known about cestodes for thousands of years, and in classical antiquity, various scholars cogitated on the nature of these organisms. It is only in more recent times that we come to understand the natural history of these animals, an understanding that has been accompanied by wonder and disgust in equal measure: wonder at the complexity and elegance of their lifecycle and disgust at how they damage the health of us and our animals.

All the 3,500 known tapeworm species are endoparasites of vertebrates, with the adult worms taking up residence in the host's gut. They are considered to be the most evolutionarily diverse of all the parasitic flatworms, with a distinctive appearance, and ranging in size from less than 1 millimeter to the enormous sperm whale tapeworm, *Hexagonoporus physeteris*, which at around 30 meters long is probably the longest invertebrate on the planet. At the head end of the tapeworm is a complex structure known as the scolex that bears a number of suckers, hooks, and spines for attachment to the intestinal wall. Behind the scolex is the tapeworm's neck, which gives rise to the largest part of many tapeworms—the strobila—a structure, actually a sequence of identical structures, unique to these animals.

The strobila is devoted to reproduction and each identical unit (proglottid) contains at least one set of male and female gonads. The tapeworms are hermaphrodites. The proglottids can fertilize themselves, exchange sperm with other proglottids, and even swap sperm with the proglottids of conspecifics. The proglottid at the back end of the tapeworm is mature, brimming with a cargo of egg capsules, and when the time is right the whole segment breaks off to be carried to the outside world in the host's feces. Like many of the endoparasitic flatworms, adult tapeworms can live for a very long time, perhaps as much as 30 years and beyond in some

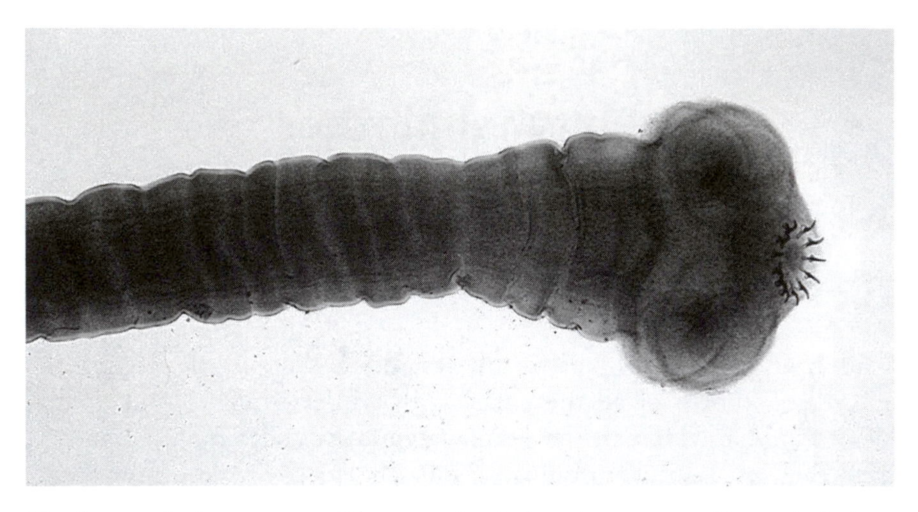

The front end of a tapeworm. These parasites infect a huge range of hosts, including humans. (CDC/Dr. Mae Melvin)

species. During their time in the host, the adult worms do relatively little damage. They are completely gutless, not constitutionally, but anatomically, and they absorb all the nutrients they need directly through their skin. However, heavy infections can cause diarrhea and immune system reactions to the waste products produced by the worm.

The tapeworms can be a problem in human and animal health because of their complex life cycle and their developmental requirements as juveniles. Typically, a tapeworm requires an intermediate and definitive host. The egg capsules in the feces of the definitive vertebrate host are inadvertently ingested by another animal—the intermediate host (a vertebrate or arthropod). In the body of this intermediate host the eggs hatch and the tapeworm juveniles burrow out of the digestive tract into the host's circulation. The aim of the juvenile worms in the body of the intermediate host is to reach the striated muscle, where they become encysted. It is these tapeworm-containing cysts that are inadvertently eaten when the intermediate host falls prey to a predator. It is these cysts that are also the most medically important phase in the tapeworm's life cycle as they can form huge cysts in organs throughout the body of the host, some of which can hold many liters of fluid. These growths can cause serious illness and even death in humans and domesticated animals. Because of the rather haphazard way in which the eggs of the adult tapeworm are scattered in the host's feces, the chances of any of the developing juveniles finding their way into the body of the intermediate host and from there into the definitive host are very slim indeed, so as an insurance the adult worms produce prodigious quantities of eggs. For example, *Hexagonoporus physeteris,* the massive tapeworm of sperm whales, has a strobila composed of at least 45,000 proglottids. Each one of these contains 4–14 sets of male and female gonads, so with such a superabundance of reproductive machinery as many as 50,000 eggs can be produced every day throughout the worm's lifetime.

Of all the known tapeworm species, only a handful are considered to be of medical or veterinary importance and we'll look at some of these in more detail below.

PORK TAPEWORM

As its common name suggests, the pork tapeworm (*Taenia solium*) requires a porcine host for part of its life cycle. The definitive hosts are carnivorous, terrestrial mammals, including humans. An infected definitive host deposits egg-laden feces on the ground that are inadvertently or intentionally consumed by a pig. The eggs hatch and the juvenile tapeworms burrow through the intestinal wall to encyst in the striated muscle, forming the stage known as the cysticercus—sometimes referred to as bladder worms. When raw or undercooked pork is consumed these bladder worms in their cysts are inadvertently ingested by the definitive host and the cysticerci go on to develop into adult pork tapeworms. This standard infection cycle does not cause significant damage to the definitive host and in many cases a pork tapeworm infection may be completely asymptomatic. However, if a person consumes food or water tainted with pork tapeworm eggs, the end result can be much more grisly.

In this scenario the eggs hatch and the infected person acts as an intermediate host to the juvenile tapeworms. In this incorrect host, the juvenile worms go onto encyst in every organ and tissue in the body, causing untold damage—a condition that is broadly known as cysticercosis. The cysticerci are most commonly found in the subcutaneous connective tissue, followed by the eye, brain, muscles, heart, liver, lungs, and body cavity. These cysts can be up to 20 centimeters across and contain 60 milliliters of fluid. It is no surprise that such a large structure aggressively growing in any part of the body can be very destructive indeed. Cysticerci in the eye can lead to retinal damage and blindness, whereas those in the brain can cause a range of unpleasant symptoms, including sudden-onset epilepsy, paralysis, and death of the brain tissue itself. In some cases, the cysticercus may die or the cyst enveloping it may be ruptured, but in both cases the end result is similar as the fluids from the cyst leak out into the host's body, causing a fatal immune reaction.

In some parts of the world, particularly central Asia, the Near East, and central and eastern Africa, the pork tapeworm is a very serious problem mainly due to the huge impact that inadvertent infection with juveniles of this species can have on human health. In the United States, only around 1,000 cases of tapeworm infection (*T. solium* or *T. saginata*) occur each year (mostly in immigrants), but in some regions of Mexico as much as 3.6 percent of the population is infected with these parasites. Worldwide, it is estimated that as many as 50 million people are infected with pork or beef tapeworm and of these at least 50,000 die each year, mainly from the complications of cysticercosis, which is more likely to occur in cases of pork tapeworm infections. The potential of *T. solium* to cause disease and death in humans is perhaps one of the reasons why pork is eschewed in many cultures or is only consumed after careful, ritualized preparation.

The impact of the pork tapeworm on human health is one thing, but it is also important to remember the impact of these animals on the health of pets and livestock. These animals can become infected in the same way as humans, harboring both the adult worms and the destructive juveniles that cause cysticercosis.

BEEF TAPEWORM

Biologically, the beef tapeworm (*Taenia saginata*) is very similar to the preceding species, but instead of a porcine intermediate host, this tapeworm takes advantage of bovines. The beef tapeworm is also one of the largest human parasites, with the adults reaching lengths of 20 meters, although 3–5 meters is more normal, much of which is the lengthy strobila, composed of as many as 2,000 proglottids.

Cysticercosis caused by beef tapeworm is rare, so the main problem this parasite incurs for human health is its competition with the host for nutrients and the numerous effects of having a large worm in the intestine: abdominal pain, diarrhea, nausea, loss of appetite, intestinal obstruction, and allergic reactions to the worm's waste products.

TISSUE TAPEWORMS

The common *Echinococcus* species (*E. granulosus* and *E. multilocularis*) tapeworms differ from the previous two species of cestode because they use humans as intermediate hosts, rather than definitive hosts. The definitive hosts for these species are carnivorous mammals, particularly canids. Also, these are small tapeworms, with adults measuring one to six millimeters long and typically trailing three proglottids. The life cycle is similar to that of the pork and beef tapeworms, but the juveniles that hatch from the eggs in the intermediate host encyst and form bladder worms in the liver and lungs. The bladder worms of this species are known as hydatids and the disease they cause, hydatidosis, can be very serious indeed. The immature stages of both *Echinococcus* species form slow-growing capsules capable of producing many juvenile worms that will go on to infect the definitive host. In *E. granulosus,* the capsules are large, multilayered cysts, but in *E. multilocularis* the hydatid has a thin outer wall and it invades surrounding tissues like a cancer, forming small pockets. When humans are infected with *E. multilocularis,* pieces of the cyst may break off and be transported to other parts of the body, where they continue to grow.

The hydatid is so slow-growing that it can take as many as 20 years for symptoms to develop following the initial infection. If the hydatid is located in the central nervous system of the host, paralysis, seizures, and blindness can ensue, whereas hydatids in the bone marrow may eventually grow to such a size that the surrounding bone thins and eventually breaks. In regions of the host's body where the growth of the hydatid is not restricted, such as the abdominal cavity, the parasite's capsule becomes enormous, holding as much as 15 liters of fluid and many millions of immature worms. Should a hydatid be ruptured, the leakage of waste products and other worm-related material into the body can be enough to cause almost instantaneous unconsciousness and death.

Of these two species, *E. granulosus* is probably more common, but *E. multilocularis* is more difficult to eradicate because of its high prevalence in wild animals. In some areas, at least 40 percent of the wild carnivore population, especially foxes, are infected with this species. In some regions of the world, such as the Peruvian Andes, the prevalence of this parasite among livestock such as sheep can be as high as 87 percent, while as many as 9 percent of the human population can be infected.

The global economic losses traceable to all the cestodes that cause disease in humans, livestock, and pets must be enormous. Each year, billions of dollars are spent on preventing humans and animals from being infected with these worms, and in cases where these parasites slip through this preventative net there are substantial treatment costs to consider as well as the losses sustained by farmers and meat producers. As is always the way in parasitology, it is the people with the most to lose who are most heavily affected by cestodes: people in developing countries who are living well below the poverty line. These people lack basic education and sanitation as well as the means of paying for drugs to prevent and treat tapeworm infections. It is these extremely poor communities where parasite infections, poor nutrition, and primitive infrastructure collude to produce a perfect storm of morbidity and mortality.

The sad fact is that tapeworm infections are remarkably easy to prevent. In most cases, infections are picked up from tainted food and water and from handling infected domestic animals. A basic knowledge of the importance of hygiene is all that is needed to break the cycle of infection and eradicate these parasites from a community. Even in situations where sanitation is lacking, tainted meat can be made safe by ensuring that it is properly cooked and that once thoroughly cooked it does not come into contact with raw, possibly contaminated food. In some of the more affluent parts of the world the fashion among well-heeled gastronomes for barely cooked beef may allow the resurgence of parasites like the beef tapeworm, creatures that for a long time have been nothing more than an agricultural pest.

FURTHER READING

Roberts, L. S., and J. Janovy, Jr. *Foundations of Parasitology.* McGraw-Hill Higher Education, New York, 2008.

Ruppert, E. E., and R. D. Barnes. *Invertebrate Zoology* (6th ed.). Saunders College Publishing, Fort Worth, TX, 1994.

Monogeneans

Once, these flukes were classified with the trematodes (see next entry), but now they are considered to be a separate class of flatworm, probably more closely related to the cestodes (the tapeworms), the other major class of flatworm parasites (see previous entry). Superficially, the monogenea resemble some of the trematodes with the addition of a complex sucker-like organ at the end of their body, which they use to fix themselves to their hosts. Unlike the trematodes and cestodes, the monogenea are predominantly ectoparasites that live on the skin or gills of fish, although there are small numbers of species that live inside certain reptiles and amphibians.

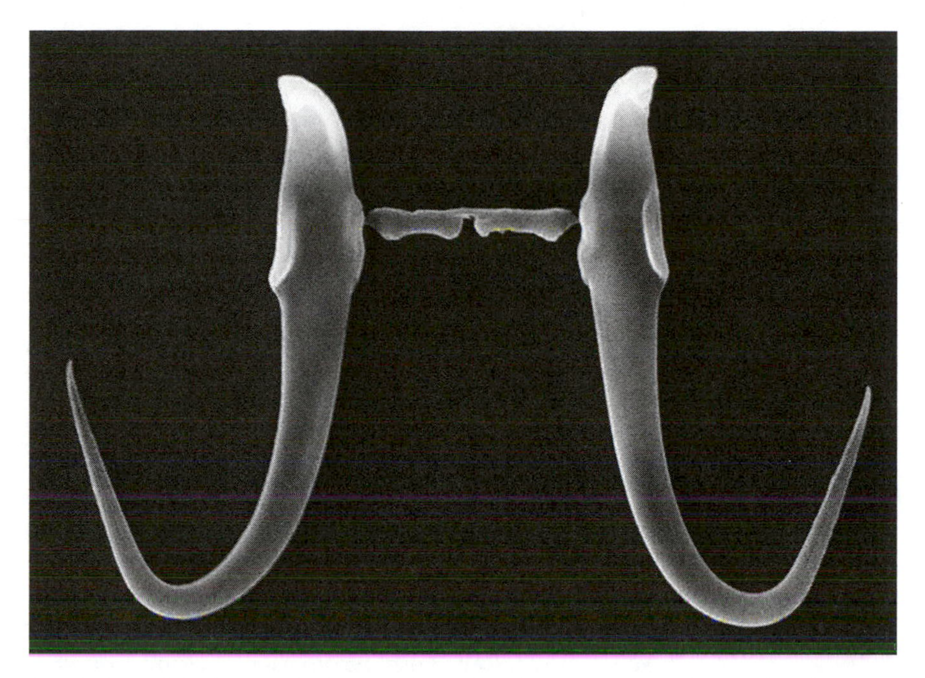

SEM plate of the haptoral sclerites of *Gyrodactylus notatae n. sp.* (in press) infecting the Atlantic silverside, Menidia menidia. (CDC)

Only one species is known from mammals and its preferred habitat as an adult is the eye of the hippopotamus.

Mainly rather small animals, the monogenea, like all flatworms, display a considerable level of internal complexity, which in the vast majority of cases can only be appreciated with the aid of a microscope. Unlike the other parasitic flatworms, the life cycle of monogeneans is relatively straightforward. The majority of species have only one host, hence their scientific name, and the typical life cycle is egg, oncomiracidium larva, and adult. This relatively simple life cycle is compounded by the fact that these animals are hermaphrodites. In each species the individuals mate, often uniting to form a long-lasting pair, facilitating the exchange of eggs and sperm. Several thousand species of monogenean are known, but being rather small, predominantly fish parasites, very little is known about them, a problem made more complex by the fact they are completely dependent on their host, so when it dies they fall off. Several thousand species of monogenean are known, but the huge diversity of fish species in marine and freshwater ecosystems suggests that a huge number of these parasites may still be unknown to science.

These worms are of no medical importance, but they can cause economic losses in fisheries. In wild populations of fish, the impact of these parasites is negligible at most, but in situations where fish are cultivated in high population densities these parasitic worms can have a devastating impact. Perhaps the most economically important genera within the monogenea is *Gyrodactylus,* which contains at least 400 species. Between them, the species in the genus parasitize a huge range of marine and freshwater fish, as well as certain amphibians. Uniquely among the monogenea, *Gyrodactylus* species adults give birth to a live young, but bizarrely, the young is actually the adult worm's sibling. Yet more remarkable is the fact that this newborn contains a further developing sibling and inside this is a fourth sibling in an arrangement akin to a set of Russian dolls. It seems these four siblings develop from the same egg, which go onto form a nested sequence of development within one another. Once the adult has given birth to its sibling, it can go about swapping sperm with another of its species and fertilize its own egg, thus allowing the process to continue. It takes around one day for these worms to mature after birth, so this remarkable reproductive strategy allows huge populations to build up very quickly on the host.

Of the 400 or so known species in this genus, it is *G. salaris* that is responsible for causing perhaps the most damage. This small (0.5–1 millimeter) worm is a parasite of many fish in both fresh and salt water, including Atlantic salmon (*Salmo salar*), rainbow trout (*Oncorhynchus mykiss*), Arctic char (*Salvelinus alpinus*), North American brook trout (*S. fontinalis*), grayling (*Thymallus thymallus*), North American lake trout (*Salvelinus namaycush*), and brown trout (*Salmo trutta*). Its native geographical range includes many of the areas that drain into the Baltic Sea, such as Finland, Russia, and eastern Sweden. For the salmon farming business this parasite can be a very destructive pest, wiping out entire stocks of these animals because of the heavy infestations that develop in these closed environments. In fish farms, the close proximity of the fishes to one another allows the parasite to spread through the captive population very quickly. In the 1970s, this species was accidentally introduced to the salmon fisheries of Norway and since that time at least 41 Norwegian rivers have been infected with this parasite, effectively exterminating the salmon populations in each. The worms themselves seem to feed on nothing more than fish mucus, some skin cells, and very occasionally blood, but the worm's attachment organ can damage the fish's skin. In heavy infestations (many thousand worms per fish)—which can develop quickly—it is this mechanical damage inflicted on the fish's protective integument that is the greatest problem caused by these worms. Thousands of tiny worms rasping at the fish's skin to gain purchase eventually cause large wounds permitting the entry of pathogens, exacerbating the decline in the health of the fish.

It has been estimated that by the mid-1980s, the introduction of *G. salaris* into Norway had caused the loss of around 300 tonnes of Atlantic salmon. The annual catch of salmon from Norwegian rivers infected with this parasite is reduced by around 90 percent, which translates as a total annual loss of 45 tonnes of salmon for Norway as a whole, with an approximate value of $27 million. If the parasite were ever to make it across the North Sea, the consequences for the Scottish salmon farming industry could be enormous, as many thousands of people are employed in an industry that contributes considerable funds to the coffers of this part of the United Kingdom. In a worst-case scenario, the introduction of *G. salaris* into Scottish salmon fisheries could result in annual economic losses amounting to around one billion dollars.

Controlling this parasite and other monogenean pests of fish is far from easy. The traditional strategy is decanting the pesticide, rotenone, into infested water, but this is a rather drastic approach, with scant evidence supporting its success in areas where it has been used. Broad-spectrum pesticides such as rotenone can kill huge numbers of nontarget organisms in aquatic ecosystems, where they are quickly dispersed by the water. This loss of biodiversity can destabilize the ecosystem, completely changing the habitats and making them unfit for the species we perceive to be valuable. A further problem with using these pesticides, especially in fish farms, is that virulent strains of pathogens and parasites may arise that can easily escape from farms and wreak havoc in wild populations.

FURTHER READING

Roberts, L. S., and J. Janovy, Jr. *Foundations of Parasitology.* McGraw-Hill Higher Education, New York, 2008.

Ruppert, E. E., and R. D. Barnes. *Invertebrate Zoology* (6th ed.). Saunders College Publishing, Fort Worth, TX, 1994.

Trematodes

Trematodes, commonly known as flukes, have among the most diverse and bizarre life histories of all the animals on earth. The 24,000 or so

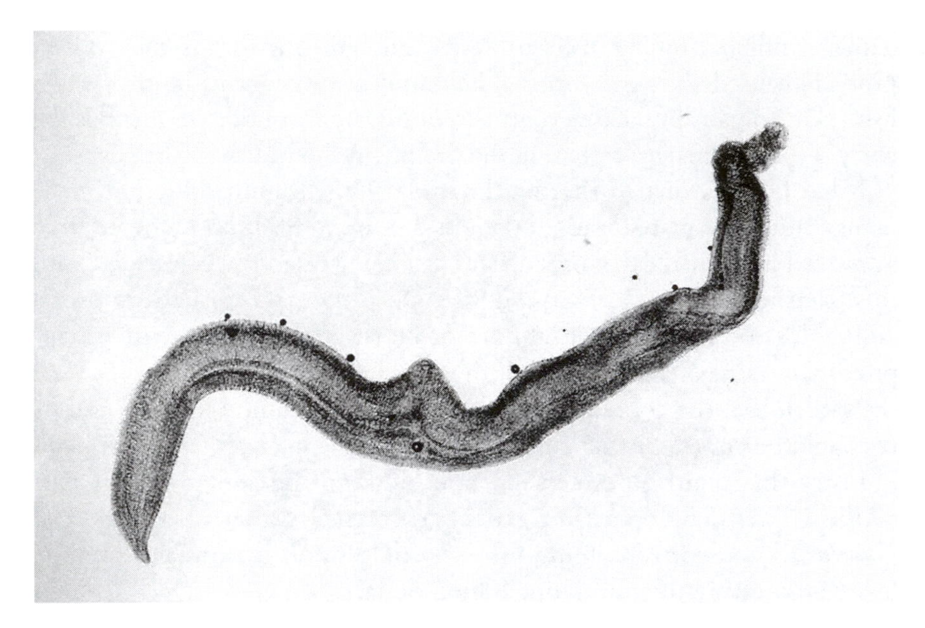

This magnified view reveals a male *Schistosoma mansoni* trematode. (CDC/Dr. Shirley Madison)

known species of trematode are internal and external parasites, taking advantage of other animals, normally vertebrates, in every habitat on Earth. However, as they're rather small, poorly known animals spending almost their entire lives in or trying to get into their hosts, their true diversity is probably greatly underestimated.

They are divided into the aspidobothrea and the digenea. None of the 100 or so known species of aspidobothrea are of any medical or economic importance, but they have attracted a considerable amount of interest from parasitologists, many of whom believe these unusual animals represent a step between free-living and parasitic organisms. The known Aspidobothrea are parasites of molluscs, fish, and turtles, but the ability of many of these species to survive outside of their host for extended periods in nothing more than salt water suggests they are only loosely tied to the hosts in which they are found. For the rest of this section we will look at the digenea, as it is this group of trematodes that contains many species of medical, veterinary, and economic importance.

The digenea are very successful parasitic organisms that inhabit the bodies of vertebrates. Indeed, there are probably very few if any vertebrates that are not parasitized by at least one type of digenean. Furthermore, every type of vertebrate organ is inhabited by the juveniles or adults of these bizarre trematodes. Their development requires at least two hosts. The first host is almost always a mollusc, but a small number of digenea use annelid worms as their first host. During the course of its development, a typical digenean goes through a number of life stages beginning with a ciliated, free-swimming larva (miracidium) that hatches from the egg. This microscopic larva penetrates the body of its first host, most often a snail, and it metamorphoses into a peculiar sac-like form known as the sporocyst.

This is where the life history of the digenean gets complicated, because within the sporocyst a number of embryos (rediae) develop asexually. The redia is slightly more complex than the previous two forms and is equipped with a gut and other structures. Like a miniature set of parasitic Russian dolls, development of the digenean gets more complicated still with yet more embryos forming in each of the redia, which will become cercariae. It is these cercariae that are considered to be the true digenean juveniles and they emerge from their snail host, often bearing a tail to aid with swimming. In all the digenea, except the blood flukes, the cercaria only becomes infective to the definitive host (a vertebrate) after it goes through one last transformation, giving rise to a metacercaria, a quiescent stage that forms a cyst in the open or in an intermediate host. In almost all the digenea it is these metacercariae that are infective to the definitive host, the animal in which the digenean reaches adulthood and reproduces.

Over millions of years, the evolution of the relationship between these parasites and their hosts has led to the emergence of some incredible behaviors, ensuring the juvenile digenean reaches adulthood. For example, the digenean *Leucochloridium paradoxum* uses various species of woodland bird as its definitive host, but to complete its life cycle many of its developmental stages are completed in a snail. Getting from the snail back to a bird to reach adulthood and reproduce presents some seemingly insurmountable challenges to the parasite. The digenean needs its snail vehicle to be eaten by a bird, but snails are wary animals that tend to stay out of sight during the day. Over the eons, the digenean has evolved a means of manipulating the behavior of its snail host and attracting the attention of birds. Like an automaton, the snail is forced into the open by its parasite, which completes its ruse by attracting the attention of the snail's sharp-eyed predators by swelling and pulsating in the snail's eyestalks. These pulsating, green-striped beacons have the desired effect and the snail is spotted and promptly eaten by a bird. This small digenean accomplishes this feat of transmission by manipulating the behavior of a mollusc and taking advantage of the predatory instincts of a vertebrate. This is just one example of the bewildering ways in which these parasites complete their lifecycle. The digenea are so diverse, yet so poorly known, we can only begin to imagine some of the relationships that exist between these parasites and their hosts. Many digenea are of medical and veterinary importance, some of which are presented in more detail below.

BLOOD FLUKES

The most medically important digenea are three species in the genus *Schistosoma (S. mansoni, S. japonicum and S. haematobium),* commonly known as blood flukes and the causative agents of schistosomiasis (bilharzia). The adults of these extremely sexually dimorphic parasites are 10–20 millimeters long. Females are longer and much thinner than the males and the two are often found linked together, with the female snugly enveloped by a large groove that runs along the underside of her mate. The miracidium larvae of *Schistosoma* species infect various aquatic snails—the intermediate host—and the cercariae that emerge from the mollusc are the stage that is infective to humans. These juveniles burrow through the human's skin in as little as 10 seconds using vigorous wiggling and digestive secretions and they continue going until they reach the peripheral circulation. Once in the blood the parasites are swept to the heart and from here they undertake complex migrations that eventually see them taking up residence, maturing, and reproducing in various locations throughout the body's central circulatory system. *S. mansoni* prefers the veins of the large intestine, *S haematobium* takes up residence in the veins of the urinary bladder, and *S. japonicum* prefers the veins of the small intestine. Like many trematodes, adult *Schistosoma* species are exceptionally long-lived, surviving for 20–30 years in their human host, but unlike other parasitic infections, it is the eggs rather than the adults of these digenea that cause disease in humans and other vertebrates. The eggs need to pass from the blood into the intestine (*S. mansoni* and *S. japonicum*) or into the urinary bladder (*S. haemaotobium*) to be voided from the body in the feces and urine, respectively. This passage of the eggs from the blood triggers immune responses and many eggs are also swept to distant parts of the body such as the brain, liver, and other organs. It is the response of the immune system to these eggs as well as the blockages that large numbers of eggs cause in various organs that we recognize as the symptoms of schistosomiasis. These symptoms include abdominal pain, diarrhea, blood in the feces, and urine, liver, and spleen enlargement, bladder and ureter fibrosis, kidney damage, possibly even bladder cancer. Neurological symptoms such as paralysis have also been observed.

Schistosomiasis is a vitally important parasitic disease. According to the World Health Organization, at least 207 million are infected with *Schistosoma* species worldwide (85% of those infected live in Africa). A further 700 million people in 74 countries around the world are at risk from this disease, the vast majority of whom live in very poor communities without clean drinking water or adequate sanitation. Mortality from schistosomiasis is very low, but morbidity is very high on account of the debilitating effects of the infection. The inability to work caused by the symptoms of this disease can have devastating consequences for whole communities in countries where it is endemic. Thirty-eight years ago it was estimated that in Africa alone, the annual economic losses from complete and partial schistosomiasis related disability was on the order of $445 million. Back then the disease was known to infect around 124 million people, so when we consider the current incidence and factor in inflation, the global economic burden of this disease today must be enormous.

LIVER FLUKES

The schistosomes are widespread parasites of humans, but there are many other digenea that cause disease in humans and livestock alike. Perhaps the most well known of these, and probably the most well known of all endoparasites, are the liver flukes (*Fasciola* spp. and *Fascioloides* spp.), large, flat leaf-shaped organisms, which spend their adult life in the bile ducts of mammals, including humans. The liver flukes find their way into the mammal definitive host by using an aquatic snail as an intermediate host and then encysting on aquatic vegetation relished by mammals. The liver flukes are a rare parasite of humans today, although the number of people becoming infected with these parasites has been steadily increasing since 1980. Over a 25-year period from 1973 to 1998, there were 7,071 cases of *Fasciola hepatica* infection in 51 countries. This is relatively minor compared to their impact on the various herbivorous mammals we have domesticated. In both humans and ungulates, liver flukes cause disease (fascioliasis) by damaging the liver, bile duct, and gall bladder. Rarely, a liver fluke infection can be fatal, but more often it is just the overall vitality of the host that is reduced as an infected animal fails to thrive. In some areas, the incidence of liver fluke infection may be as high as 70 percent, which carries a significant economic burden for livestock farmers attempting to rear healthy animals for meat and milk production.

Closely related to the flukes discussed above is *Fasciolopsis buski,* another large trematode that spends its adult life in the small intestines of humans and pigs, where it can cause disease by triggering immune responses, blocking the intestine, and physically damaging the delicate wall of this organ. In some cases, infection with this trematode can be fatal. The infection cycle depends on night soil (manure containing human feces) being used as crop fertilizer. It was estimated back in the 1940s that at least 10 million people were infected with this parasite, mostly in India, China, and Southeast Asia. If this estimate was accurate, we can be certain this trematode currently infects even more people in some of the most densely populated regions on Earth.

Two further genera of fluke that are found in and around the liver and which are considered to be of medical importance are *Clonorchis* and *Opisthorchis. Clonorchis sinensis* is widespread throughout Japan, Korea, China, Taiwan, and Vietnam, where it finds its way into the human population via the consumption of dried, pickled, or frozen fish. The adults of this species live in the bile ducts of humans and large numbers of these little creatures can cause considerable damage to this structure and also to the liver. Typically, anywhere between 20 and 200 of these flukes have been found in the bile duct of an infected person, but in exceptional cases around 20,000 adults have been removed from a single human. Worldwide, several million people are probably infected with this trematode. Two species of *Opisthorchis* are of medical importance—*O. felineus,* which has a life history similar to *C. sinensis,* but is more commonly encountered in Europe, and *O. viverrini,* a parasite known from Southeast Asia with a high prevalence in northeast Thailand. Both species cause disease in the same way as *C. sinensis,* but only *O. viverrini* is of comparable public health importance. *O. felineus* probably infects at least one million people around the globe, while the incidence of *O. viverrini* is slightly higher.

LANCET FLUKE

The lancet fluke (*Dicrocoelium dendriticum*) is another important parasite of domesticated ungulates. Like the liver flukes it too lives out its adult life in the bile ducts of these definitive hosts. Unlike the liver flukes, the lancet fluke has dispensed with an aquatic stage, instead using various species of terrestrial snail as its intermediate host. The symptoms of a lancet fluke infection (dicrocoeliasis) are essentially the same as fascioliasis; although huge numbers of these trematodes can be present in a single definitive host (more than 50,000 of these parasites have been found in a single sheep). This trematode is known from at least 31 countries and in some areas its prevalence may be high as 70 percent.

LUNG FLUKES

Digenea in the genus *Paragonimus,* commonly known as lung flukes, are also important parasites of carnivorous mammals, including humans. Two species, *P. africanus* and *P. uterobilateralis,* are considered to be the most important parasites of humans in this genus. Two intermediate hosts are required by these parasites—a freshwater snail and a freshwater crustacean, such as a crab or crayfish. When a mammal eats one of these crustaceans, the mammal becomes infected and the juvenile parasites first encyst in the duodenum of the small intestine before piercing the intestinal wall and embedding themselves in the abdominal wall. Several days later they reenter the body cavity and eventually find their way to the lungs where they mature. It is during the course of these wanderings that juveniles can cause disease by lodging in various organs of the body and triggering immune responses. In cases where large numbers of juveniles inadvertently find themselves in the central nervous system and heart, death may even result. Worldwide, at least 21 million people are thought to be infected with this trematode, mostly in Southeast Asia, Africa, and South America. The normal route of human infection with this parasite is via the consumption of raw or undercooked freshwater crustaceans.

As we have seen, trematodes are extremely important parasites of both humans and our domesticated animals, causing debilitating disease, death, and huge economic losses. In view of the impact of these animals it comes as no surprise that scientists around the globe are trying to develop ways of eradicating these trematodes, particularly the *Schistosoma* species, which are solely responsible for abject misery and the perpetuation of poverty in many subtropical and tropical regions. Try as we might, the fight against these parasites is far from being won and in many cases, the looming specter of drug resistance means that many of the drugs routinely used to control infections of these trematodes are becoming less effective with every passing year. The key to controlling these parasites lies in unraveling the complexities of transmission and breaking the cycle of infection.

The life cycle of these animals has been well studied and for most species of medical importance we know the hosts they require to complete their development. Understanding the biology of these intermediate hosts is also crucially important in identifying those parts of the infection cycle that can most easily be broken.

In the case of schistosomiasis we know that infection by the trematodes that cause this disease and poverty go hand in hand. The vast majority of people with schistosomiasis do not have access to clean water and are therefore forced to bathe and wash in whatever sources of water are available to them—normally pools and lakes infested with the juveniles of these parasites. Crucially, it is the lack of even simple sanitation that keeps the cycle of infection intact. The use of very simple latrines and rudimentary education in the importance of basic hygiene would ensure that the eggs of *S. mansoni, S. haemtobium,* and *S. japonicum* never found their way into water that supported the intermediate molluscan hosts, thus breaking the cycle of infection. However, this problem goes even further, because in many ways this poverty is a direct result of too many people trying to eke out an existence on too few resources. Until we address the problem of the burgeoning human population, lots more people will live short, unproductive lives plagued by parasites such as these.

On a final, more positive note, there is considerable evidence to suggest that trematode infections may be a good thing in that they reduce the incidence of allergies and autoimmune disorders in those populations where they are endemic, more of which is explained in the introduction.

FURTHER READING

Roberts, L. S., and J. Janovy, Jr. *Foundations of Parasitology.* McGraw-Hill Higher Education, New York, 2008.

Ruppert, E. E., and R. D. Barnes. *Invertebrate Zoology* (6th ed.). Saunders College Publishing, Fort Worth, TX, 1994.

Vertebrates

Black Rat

Rattus rattus, commonly known as the black rat, ship rat, roof rat, and house rat, is a very common animal. Like its very close relative the brown rat (see next entry), the black rat has hitched rides all over the planet and today it is one of the most widely distributed rodents. The black rat is thought to be a native of tropical Asia, possibly Indo-Malaysia. As human settlements sprang up in this part of the world and ocean-going trade established the first links with distant cities, the black rat was quick to monopolize and take advantage of the opportunities offered by human activities.

The ancestors of today's black rats were probably tree-dwelling coastal animals in their native range and they have probably associated with humans for many thousands of years. Their preference for slightly warmer climates means that today, in many temperate areas, the black rat has been out-competed and edged out by the brown rat, which is better suited to cooler climates. However, in tropical and subtropical areas, the black rat is the most frequently encountered rat species, particularly in coastal locations. Like other pest rodents, the black rat is a prodigious breeder and in many locations will reproduce throughout the year. Gestation lasts 21–29 days with the female giving birth to an average of seven young. These offspring are independent by around 3–4 weeks old and are sexually mature at 12–16 weeks. A single female black rat can produce 3–5 litters in a single year and in no time at all a population founded by a single, pregnant female rat can be very large indeed. Interestingly, black rats live in small groups known as packs numbering as many as 60 animals. These packs consist of several males and one or two dominant females. Among the males in the group there is a hierarchy, with the more dominant males having the opportunity to mate with the females. As with almost all small rodents, the exceptional fecundity of the black rat in the wild is balanced by a short life span and very high mortality from predators, parasites, and pathogens. In human settlements, the rat's natural predators are rare, food is very abundant, and their populations explode accordingly.

Engraving from 1500 of a plague victim in bed pointing out to three physicians the swelling or boil under his armpit. (Library of Congress)

Apart from the obvious differences in appearance between the black and brown rats, the former species is a very good climber thanks to its tree-dwelling heritage. In and around human settlements the black rat is often found high up in buildings, hence one of its common names—the roof rat. Its ability to scale a range of surfaces makes preventing the entry of black rats into buildings very difficult. A further divergence in the natural history of the black and brown rats is their respective diets. The brown rat is a committed omnivore with very catholic tastes, while the black is more selective, preferring vegetable food, such as fruit, grain, and cereals. When these foodstuffs are in short supply it will also eat invertebrates, small vertebrates, refuse, carrion, and feces.

A more selective feeder than the brown rat, the black rat is still a major pest of agriculture around the world, where it damages crops in the fields as well as consuming and tainting stored crops. A single black rat needs to

The black rat has spread around the globe with humans but has been supplanted in many areas by the brown rat. (CDC)

consume around 15 grams of food every day, so when there are millions of these animals in any given area they are capable of consuming vast quantities of food and contaminating yet more with the copious urine and feces they produce.

The black rat is not only a serious pest of agriculture. Like all rodent pests, the black rat is also of major public health importance because of the many pathogens and parasites it harbors. As already mentioned in the brown rat entry, the black rat is actually a more effective reservoir of the pathogen that causes bubonic plague compared with its other common congener. The role of the black rat in the plague of Justinian and the Black Death is open to debate, but it's possible this rodent played a role in the spread of these diseases. Aside from its involvement in the transmission of bubonic plague, the black rat is also associated with leptospirosis (Weil's disease), rat bite fever, hanta virus, Q fever, toxoplasmosis, cryptosporidiosis, murine typhus, and conditions caused by *Salmonella* species bacteria and trichinellosis. Collectively, these diseases are responsible for considerable mortality and morbidity in humans around the world, particularly in developing nations where poor infrastructure provides perfect breeding opportunities for rats, as well as the various filth-loving parasites and pathogens they spread.

The Plague

The plague is the collective name given to a series of pandemics that swept through the Old World beginning in the sixth century, two of which are very poorly understood. The terms plague, Black Death, and bubonic plague are often used interchangeably and synonymously, but it is important to note they actually refer to three distinct pandemics:

- Plague of Justinian (sixth century A.D.)
- The Black Death (14th–18th century)
- Bubonic plague (19th–20th century)

These pandemics, especially the first two, are still somewhat of a mystery and subjects of intense discussion and contention among many scholars. These experts argue about the causative agent of these pandemics, whether each outbreak was caused by the same pathogen, and what creatures were vectors and reservoirs for these pathogens.

The Black Death is the best-known of these pandemics and probably the most devastating as it ravaged Europe, North Africa, and the Near East in an episodic pandemic that lasted over 350 years from 1347, with each episode killing as many as one-third of the population. In medieval England, the Black Death is thought to have killed 30–50 percent of the country's entire population of around 4 million people between 1348 and 1350. The impact of the Black Death was so massive that society, particularly in Western Europe, was never the same again. Following the final major outbreak of the Black Death in Europe in the 18th century, the disease mysteriously disappeared as rapidly as it had emerged many years before. One possible explanation for this is the supplanting of black rats by brown rats and their respective abilities to act as reservoirs for the disease—explained in more detail below.

All we know for certain about the three pandemics listed above is that they killed huge numbers of people and that the most recent—the bubonic plague—was caused by the bacteria, *Yersinia pestis,* a rod-shaped bacterium isolated from infected individuals. The vector of this bacterium is the oriental rat flea (*Xenopsylla cheopis*), an endoparasitic insect on a variety of rodents, including *Rattus* species. These fleas, when they come into contact with humans, will bite and attempt to feed, transmitting the bacteria, while the rodent population acts as a reservoir for the bacteria. The brown rat is effectively immune to *Yersinia pestis,* limiting the ability of this rodent to act as a reservoir. The black rat, on the other hand, is very susceptible to infection by this bacterium, making it an ideal reservoir for the bubonic plague.

Other than assumptions drawn from what we know about bubonic plague, the facts surrounding the plague of Justinian and the Black Death have been lost in the mists of time It is now widely accepted that the Black Death was not caused by the bacterium responsible for bubonic plague. Instead, it has been proposed that the Black Death has many of the hallmarks of a viral hemorrhagic disease. Understanding the natural history of these two ancient pandemics is as important to contemporary global public health as modern pandemics or the threats thereof. Whatever pathogen(s) were responsible for causing the plague of Justinian and the Black Death, we can be assured there is a very high possibility they are still around in a state of quiescence. Indeed, it is well known that *Y. pestis* is still endemic in many parts of the world with many species of wild and domesticated mammals and their attendant fleas carrying this pathogen. Every year *Y. pestis* is responsible for the deaths of hundreds of people, particularly in Africa and the Indian subcontinent. This low rate of mortality may represent a dormant state for the bacterium until the collusion of chance events initiates another pandemic.

Further Reading: Orent, W. *Plague: The Mysterious Past and Terrifying Future of the World's Most Dangerous Disease.* Free Press, New York, 2004; Prentice, M. B., T. Gilbert, and A. Cooper. Was the Black Death caused by *Yersinia pestis? Lancet Infect Dis.* 4(2)(2004): 72; Byrne, J. P. *The Black Death.* Greenwood, Westport, CT, 2004; Scott, S., and C. J. Duncan. *Biology of Plagues: Evidence from Historical Populations.* Cambridge University Press, Cambridge, MA, 2001.

The opportunistic nature and incredible reproductive potential of black rats combined with their habit of stowing away on ships and disembarking at the first available landfall makes them excellent colonists, especially of isolated oceanic islands with a dearth of large terrestrial predators. The litany of extinctions of island wildlife that characterized the age of exploration is partly attributable to the introduction of foreign species by humans, purposefully or otherwise. The black rat should be considered one of the most damaging introduced species. Aboard ships it has found its way to countless islands, many of which have been isolated in space and time for millions of years. The ecological consequences of a black rat introduction are exemplified by the islands of New Zealand. The black rat was the second rat species to be introduced to these islands following the arrival of the Polynesian rat (*Rattus exulans*), which accompanied the ancestors of the Maori during the Polynesian colonization of Oceania. Since its arrival in New Zealand in the 1850s, the black rat has spread from the North to the South Island and rather than just living in and around human settlements it has adopted a more wild, sylvatic existence, surviving and thriving throughout the natural habitats of these islands. Many of the native animals of New Zealand are woefully ill-equipped to deal with an opportunistic animal like the black rat because they evolved in isolation without any native terrestrial predators trying to eat them. For the black rat, the abundance of animals with no real way of protecting themselves or their young has provided a veritable banquet and since the arrival of this pest in New Zealand several endemic animals have become locally or completely extinct, including birds, reptiles, and insects. Unfortunately, the black rats, along with the other introduced rats (the Polynesian and brown rats), are now so widespread that eradicating them is a practical impossibility. For the sake of conserving the extremely vulnerable animals that have no defenses against the army of rats, the authorities in New Zealand have decided to create rat-free reserves by eradicating these animals from many small islands surrounding the North and South island. Unless a ship runs aground or rats from the mainland somehow manage to swim across to these refuges the native New Zealand fauna will be safe. Interestingly, the impact of the introduction of black rats to New Zealand may not be completely negative. There is some evidence to suggest that New Zealand black rats, which have taken to living in the island's forests, may be dispersing the seeds and spores of various native plants and fungi. Many of New Zealand's native seed and spore dispersers became extinct long ago, so the long-term viability of the island's natural habitats may ultimately depend on introduced animals like the black rat.

Controlling the black rat can be very expensive and time-consuming. The best way of controlling this rodent is restricting its access to places where the rats forage for food and make nests. Food and refuse should never be freely available as this will attract these animals. In situations where an infestation has become established, baits and traps are the best option for controlling black rats (see brown rat entry).

Putting a dollar value to the diseases transmitted by black rats, the crops they damage, and the ecological consequences of their inadvertent introductions is impossible, but these animals must rate as one of the most costly of all pests. The involvement of black rats in bubonic plague is undeniable, a disease that caused the death of at least 12 million people in India and China alone. In today's terms, a pandemic of this scale would cost hundreds of billions of dollars in prevention, treatment, and lost productivity. The other pathogens and parasites transmitted by the black rat can cause serious illness and death, so the global economic burden of these must be huge.

The ranges of many pest rodents overlap, so attributing crop damage to just the black rat is not possible; however, these animals collectively consume and contaminate huge quantities of food every year, food intended for humans and domesticated animals. Estimating the cost of black rat introductions to isolated islands and archipelagoes is even more difficult. Putting a price on the extinction of a native species and the ecological changes that are brought about by the introduction and spread of an invasive species is impossible. All we can conclude is that the black rat is a very damaging pest, but perversely, it also deserves our admiration for its adaptability and ability to thrive no matter how hard we try and eradicate it.

FURTHER READING

Buckle, A. P., and R. H. Smith (eds.). *Rodent Pests and Their Control.* Oxford University Press, Oxford, United Kingdom, 1996.

Brown Rat

Rattus norvegicus, also known as the common rat, sewer rat, and Norway rat, among other names, is one of the most infamous animals on the planet. This small, shy, unassuming rodent has its origins on the plains of Asia, in what is now northern China. Indeed, the burrow-dwelling wild relatives of the ubiquitous brown rat can still be seen living in this area today. Biologically, the brown rat is one of the largest members of the mouse family and like most rodents it has very sensitive senses of smell and hearing. Like their smaller relatives they are also very gregarious animals, able to live quite happily at high population densities provided there is sufficient food. They are also vocal and can produce many different types of sound, many of which are beyond the range of human hearing. The most incredible feature of this animal's biology is its reproductive ability. Eighteen hours after giving birth a female brown rat is ready to mate again and the gestation period is a mere 22–24 days. With such a rapid reproductive cycle one of these female rodents is capable of giving birth seven times a year, spawning around 60 offspring. Very few other mammals approach this level of fecundity. Their staggering reproductive potential is combined with a very catholic attitude toward food. Primarily grain and seed eaters, brown rats are omnivorous, able to survive and indeed thrive on a huge range of foods. To process this food, the brown rat's jaws are very powerful for its size and their continually growing incisors are used to good effect to gnaw food while it is held in the front paws.

Unlike many pests, such as mosquitoes, it is very unlikely the brown rat has been a pest of humans for a geologically significant period of time. When we just were hunter-gatherers the only relationship that existed between us and rats was that of predator and prey as our ancestors on the Asian plains may have utilized these rodents for food. Only when humans relinquished a hunter-gatherer lifestyle in favor of a more settled, agricultural way of life did this relationship change. As soon as humans took to an agricultural way of life the evolutionary trajectory of the brown rat became inextricably bound to our own.

For an emerging agricultural society to work, a certain degree of food storage is required, both to feed a high-density population and as a stockpile for when pickings are slim. These stockpiles were a bounteous source

The brown rat has exploited the opportunities offered by human settlements and is now one of the most widespread and successful of all mammals. (AP/Wide World Photos)

of food for any animal opportunistic enough to make use of them. Several animals, mostly small rodents, began to associate themselves with the settled communities of our ancestors because of the easy pickings on offer. It is known that some plants had been domesticated in China 9,500 years ago, so rats have been associated with the human race for at least this long. Because this promising agricultural lifestyle was adopted relatively swiftly over huge swaths of Eurasia we can be sure the geographical range of the ancestors of the brown rat underwent a similar expansion.

Several characteristics of the brown rat's ancestors preadapted them to a way of life that would bring them into direct conflict with humans. Firstly, they were generalists when it came to food, allowing them to thrive on the range of crops our ancestors were growing and storing. Secondly, they were small and wary, enabling them to keep out of sight and evade danger. And lastly they were prolific breeders, an attribute enabling them to build up large populations to exploit the food resources available. Thanks to these characteristics the brown rat thrived wherever there were settled communities of humans and today it is found all over the world on every continent except Antarctica.

The brown rat is considered to be a pest for a number of reasons. Firstly, rats transmit some nasty diseases via their saliva, feces, and urine, and they harbor parasitic animals capable of passing on yet more pathogens to humans, making them reservoirs of disease. Secondly, and perhaps the most pressing problem with rats in the modern day, they impact food production. The brown rat is also responsible for untold damage as an introduced species and on isolated islands these rodents can wipe out native, often very rare animals. Finally, their burrowing and propensity for gnawing things can be very destructive, especially when they are found in high densities.

Rat-borne diseases are a significant problem in human and domesticated animal health and it is often said that rats, especially the brown and black rats, have been responsible for more deaths than all the wars and revolutions combined through the ages. This arresting statistic is not based on fact and although rat-borne diseases have undoubtedly killed millions of people over the centuries, the actual numbers are impossible to know. The most infamous disease definitively associated with rats is the bubonic plague (see sidebar in black rat entry), but it is very likely it was the black rat rather than the brown rat that was the primary reservoir in the spread of this disease. Apart from bubonic plague, rats are also associated with a number of other diseases, including leptospirosis (Weil's disease), rat bite fever, hanta virus, Q fever, toxoplasmosis, cryptosporidiosis, murine typhus, and conditions caused by *Salmonella* species bacteria.

Collectively, brown rats consume and damage huge quantities of human and domesticated animal feed every year. A single rat can get through 9–18 kilograms of food every year, which may not seem like a great deal, but when you consider there are probably billions of these rodents around the world their collective appetite is enormous. They are also known for consuming just about anything, from stored grain, sugar cane, and fruit while it's still on the tree to decaying matter in garbage dumps. There's little that isn't on the menu for the brown rat. As they gnaw and nibble, they also produce droppings and urine in copious quantities. Three hundred rats living in a grain store will produce 15,000 droppings and 3.5 liters of urine every day, not to mention countless shed hairs and greasy skin secretions, all of which contaminate huge quantities of food destined for humans and domesticated animals. It has been estimated that rats contaminate 10 times the amount of food they actually eat. In rice-growing regions of the world, especially Southeast Asia, rats along with several other rodent pests are capable of eating and spoiling a significant proportion of the annual rice harvest (see sidebar), food that could be used to help meet the dietary requirements of many millions of people.

As the brown rat is a generalist when it comes to food, live animals are also taken, including other small mammals, birds, reptiles, amphibians, and invertebrates, especially if they're of a smaller size. In most areas this isn't really a problem, but if these rodents find their way onto islands, typically on ships, they can do untold damage to the native fauna. Birds are particularly at risk from introduced rats, especially those species that have evolved in the absence of predators. Without any enemies to escape from, these birds often forsake the power of flight and build their nests on the ground. Rats that find themselves on an island of ground-nesting birds are confronted by very rich pickings and it is probable that rats have had a paw in the extinction of many unique island animals, not only birds. The same is also true of seabirds as their breeding colonies are often on islands where large terrestrial predators are absent. Should rats find their way on to these islands the bird populations can be devastated as the rodents will consume the eggs, nestlings, and even the adults of smaller species. Eradicating rats once they've been introduced is very expensive and time-consuming and there are no guarantees of success.

Disease transmission and eating aside, rats also have a propensity for burrowing and gnawing, two other activities that can lead them into direct

Rodents and the Rice Harvest

Throughout Southeast Asia and the Indian subcontinent, rice is the staple food for hundreds of millions of people. In India alone, the rice harvest in 2007 amounted to 144 million tonnes. In these areas, rodents are the most important rice pest, especially when the crop is still in the field. In Mizoram, a state in the northeast of India, the rice harvest in 2008 was around 45,000 tonnes, of which around 40,000 tonnes was consumed or damaged by rats—a loss of almost 90 percent, which affected about 70 percent of the farming families in the area.

In Indonesia, a country that produced 57 million tonnes of rice in 2007, rodents are responsible for crop losses of around 17 percent. Rice production in Vietnam in 2007 was around 35 million tonnes and of the total area in this country planted with this crop, rodents caused damage to over 700,000 hectares in 1999. In some parts of Vietnam, rodent pests are known to outnumber humans by at least 10 to 1.

Controlling the rat problem in these areas would allow for more efficient use of land and a reduced need to cultivate the natural habitats that are so crucial for biodiversity.

confrontation with humans. Burrowing can be a real problem when brown rats are present in high densities as the burrows can cause subsidence, flooding, and soil erosion, and can damage roads, buildings, earthworks, and sewers. The same can also be said of the brown rat's predilection for gnawing. Rats will gnaw just about anything, sometimes to keep the incessant growth of their incisors in check, but often simply because they are trying to eat the material in question. The plastic insulation around electrical wires is a common favorite and this seems to be gnawed because of the odor of some of the compounds in the material. Gnawing electrical wires normally ceases when the rat breaches the insulation and receives a fatal electric shock; however, the resultant short circuit and sparks are enough to start fires or disable important electrical equipment.

Rats have many things against them when it comes to winning human admirers, so it is no surprise we wage an ongoing war against these resourceful animals—a war we can never win. Rats are intelligent, adaptable, and prolific breeders. Try as we might to eradicate them, their numbers continue to swell. Their wariness makes them difficult to catch, so bait traps are the standard means of controlling a rat infestation. Bait traps consist of food or liquid that is toxic to the rat. Decades ago, these rat poisons were simply toxic compounds, potentially lethal to all animals, but over time and with lots of research, better rat poisons were created. The standard rat poisons used today are anticoagulants that prevent the rat's blood from clotting, so it bleeds to death. The first of these, warfarin, was something of a breakthrough, but the rat's incredible ability to reproduce means resistance has now started to emerge and some populations of brown rat in urban areas are completely resistant to warfarin. New, even more potent anticoagulant compounds are available, but control of the brown rat cannot focus solely on poison baits. Controlling the brown rat is more about limiting the opportunities available to them in terms of food resources and places to live.

There's no doubt the rat is a much-maligned creature, but in defense of this rodent you have to admire its opportunistic nature and its ability to thrive just about anywhere. The stinking sewers of an urban street are a long way from the plains of Asia, the ancestral home of this species, but the varied places it has managed to colonize are testament to what a successful animal this is. In terms of abundance and geographic range the brown rat is the most successful mammal on the planet after our own species. We might not like it, but we share the rat's attributes of opportunism and adaptability and when we pour scorn on these animals we should remember they are simply achieving what every species strives for: success.

FURTHER READING

Buckle, A. P., and R. H. Smith (eds.). *Rodent Pests and Their Control.* Oxford University Press, Oxford, United Kingdom, 1996.

Cane Toad

The cane toad (*Bufo marinus*) has earned a bad reputation for the damage it has done to native wildlife in many areas of the world, particularly Australia, but before we go any further let's remember that humans created the cane toad problem by purposefully introducing this animal into various countries.

This toad is a native of Central and South America. The adults are giants of the amphibian world, weighing up to 1 kilogram. They're also prolific breeders as females can produce 8,000–35,000 eggs, sometimes twice a year. In tropical areas the young grow rapidly and they can be sexually

Cane toads were intentionally introduced to Australia in 1935 and have since had a devastating effect on the native wildlife. (Rewat Wannasuk | Dreamstime.com)

mature within a year. One of the most interesting features of this toad's biology is the presence of large parotid glands on its neck that produce a potent toxin. Any potential predator that threatens the cane toad is treated to a threat display in which the toad directs its oozing glands toward the attacker. In small and medium-sized animals this toxin can be lethal and there are even reports of humans dying following exposure to cane toad toxin. All life stages of the toad are toxic, even the eggs, but a soft-bodied animal like a toad needs defenses to keep its natural enemies at bay. Interestingly, the toad's Latin name suggests it lives in sea water, but this name stems from the observations of early naturalists who mistakenly believed the toad lived in both terrestrial and marine environments. The permeable skin of an amphibian means a dip in salt water would be fatal.

How did the cane toad go from being just a big, warty amphibian to a major pest? Like so many of these poorly thought-out introductions, we have to go back to the 19th century and some enterprising gentlemen with links to the lucrative sugar cane industry, who decided in their infinite wisdom that the cane toad had promise as a biological control agent for the multitudinous pests of sugar cane. In 1844, a consignment of the toads was introduced to the island of Jamaica in the hope of controlling the burgeoning rat population that was troubling the plantation owners. This initial introduction proved unsuccessful, but it didn't prevent more toads being introduced to Puerto Rico in the early 20th century as a weapon against an outbreak of sugar cane beetles ravaging the island's valuable crop. Interestingly, this introduction was touted as a success and by the 1930s *Bufo marinus* was lauded by experts as a biological control agent of considerable potential. How very wrong they were.

No sooner had the toad been introduced to other islands in the Caribbean, the Philippines, and many islands in the Pacific, Australia, New Guinea, and Florida than it rapidly turned from farmer's friend to agricultural enemy number one. If the toads introduced to Puerto Rico had been eating sugar cane beetles they were something of the anomaly. All the other introductions played out with the toads turning their noses up at the sugar cane beetles in favor of anything else they could jam into their sizeable mouths, including all kinds of invertebrates, other amphibians, reptiles, birds, and small mammals.

The introduction of this amphibian into Australia is perhaps the most infamous of all the invasive animal sagas because of the damage it has inflicted on the native fauna. One hundred and one toads were shipped from Hawaii to North Queensland, Australia, in 1935. Although the initial release was followed by a ban to allow a study of the animals in their new environment, larger-scale releases were eventually ratified and by 1937 more than 60,000 toads had been released. By the early 1980s the toads had spread into the Northern Territory. Today they continue to march across the country at a rate of about 30–50 kilometers per year in the Northern Territory and about 5 kilometers per year in northern New South Wales. The toads are seen as a pest for the following reasons:

- Eating native animals
- Competing with native animals for food
- Transmitting diseases and parasites to native amphibians
- Poisoning pets and injuring humans with their toxins
- Poisoning native animals that prey on amphibians
- Eating honeybees

The cane toad is undoubtedly a problem in Australia, although some reports of their impact may have been exaggerated. It is thought native predators can be affected when the advancing front of cane toads moves into a new area. These native animals will have had no previous experience of this amphibian, so they'll attempt to eat it, with the end result of extreme sickness or even death. However, once the predators in an area, particularly birds and mammals, become accustomed to the toads they will learn through experience to give them a wide berth. The cane toad has been most successful in disturbed habitats where its interaction with native amphibians is probably very limited, but in more natural habitats there is a very real possibility the cane toad may transmit diseases to native frogs, toads, and newts. Of particular concern is the chytrid fungus, a pathogen seemingly responsible for devastating amphibian populations worldwide. Competition with native animals for food may be one of the most important effects of the spread of the cane toad as it's a generalist with a large appetite, thus depriving native animals with similar albeit more restrained tastes. The impacts on humans and pets are very minor compared with the wider impacts on what are decidedly sensitive ecosystems.

Eradicating the cane toad is feasible, but it would be monstrously expensive. The species is now so widespread that complete eradication would run into hundreds of millions, if not billions of dollars. About one million Australian dollars are currently spent each year on cane toad control research, although more investigations need to be made into the impacts of this species as many experts argue that the effects of the toad on native ecosystems may be negligible and that the resources directed at trying to control this animal would be better spent on controlling other invasive species or conservation projects for native flora and fauna. Research is currently identifying pathogens or native species that could be used to help control the cane toad. The meat ant (*Iridomyrmex reburrus*) has been identified as a contender because meat ants attack the young toads when they begin a life on land. Unlike native frogs and toads, young cane toads are active during the day, which puts them on the menu for the meat ant—a voracious, diurnal predator.

FURTHER READING

Lever, C. *The Cane Toad: The History and Ecology of a Successful Colonist.* Westbury Academic and Scientific Publishing, New York, 2001.

Ward-Fear, G., G. P. Brown, M. Greenlees, and R. Shine. Maladaptive traits in invasive species: In Australia, cane toads are more vulnerable to predatory ants than are native frogs. *Functional Ecology.* [Online]. Accessed March 31, 2010. http://onlinelibrary.wiley.com/doi/10.1111/j.1365-2435.2009.01556.x/pdf.

European Rabbit

The word *rabbit* conjures up images of cartoon characters and memories of fluffy pets, but in the wrong places these mammals can be an unparalleled nuisance. Before we start demonizing these undeniably cute animals, let us not forget that the rabbits themselves are blameless because it is humans who have transported them to areas where they were never naturally found, in much the same way as the cane toad.

The story of the European rabbit begins in the Iberian Peninsula and North Africa, the natural home of this species. Meddling with the distribution of animals is not a new thing—it has been going on for millennia and among the menagerie of the Romans was the rabbit. After conquering the Iberian Peninsula and North Africa they saw the rabbit as a useful species—it was an unfussy eater and a prolific breeder, plus it had lean,

The European rabbit has been introduced into many areas around the world, with disastrous consequences for native flora and fauna. (Edurivero | Dreamstime.com)

nutritious flesh and thick fur. The Romans were swift to adopt anything of use in the areas they conquered and they transported rabbits to their colonies throughout Western Europe. Much of the Roman Empire was a far cry from the arid, rugged land of the southern Mediterranean, and with so much lush vegetation on offer, the rabbit thrived. By the 12th century the rabbit had reached the United Kingdom and over the centuries and decades its population swelled massively. Rabbits have undoubtedly caused damage to habitats and crops in historic times, but as there is no relevant documentation we cannot be sure how much.

Rabbits have been at their most devastating in Australia. The first British fleet traveling to Australia had rabbits aboard in 1788 and when they made landfall the rabbits took up residence in these distant lands. By as early as 1827 feral rabbits were abundant in Tasmania, but their relatives on the mainland were restricted to the area around Sydney. All this changed in 1859 when Thomas Austin released about a dozen of the animals on his land in Victoria for the purposes of hunting. Hunt them he did, but the rabbits found themselves in a land similar to their ancestral lands and they bred feverishly, so much so, that by 1910 they had spread to include nearly all of their present-day range, which is most of Australia apart from the northernmost areas.

Rabbits are a real problem in Australia because they damage crops and native flora, compete with livestock and native fauna, and cause erosion from their burrowing. The annual economic damage incurred by rabbits is estimated at $600 billion. These animals are able to cause so much damage because they are very unfussy eaters and they breed so rapidly. A female rabbit is sexually mature at five months of age and in a single year she can produce four to seven litters of anywhere between 2 and 12 young. This means a single female rabbit can easily produce 40 young every year. With such high fecundity it is not unknown for a rabbit population to multiply 8–10-fold in a single breeding season.

These lagomorphs have been such a bane for Australia that there's been a lot of interest in getting rid of them for a long time. As early as 1919 the curiosity of scientists was aroused by a *Myxoma* virus that affects South American cottontail rabbits, causing a small, benign lump. Their hope was this virus could be used as a biological control agent to stem the populations of rabbits in Australia, so investigations were initiated to assess the pathogen's potential. The virus only infected lagomorphs and in contrast to the cottontail rabbits in which it was naturally found, it was lethal to the European rabbit. Research was continued to gain a better understanding of the virus. By the late 1940s there was real hope that this pathogen and the disease it caused, myxomatosis, could be used to deal a heavy blow to the burgeoning rabbit population. In the summer of 1950 a strain of the *Myxoma* virus was introduced into the feral rabbit population in the Murray River valley and in the following two years it was transmitted by mosquitoes throughout all the rabbit-infested areas in Australia. The short-term results of this release were amazing: 99.8 percent of infected rabbits died and the rabbit population in Australia fell by 95 percent. This massive decline meant that sheep no longer had to compete for food with huge numbers of rabbits and so wool production boomed, allowing an additional 32 million kilograms to be produced in 1953. Following the release of the virus in Australia is wasn't long before it was released, legally or otherwise, in other countries with rabbit problems. It appeared in France in 1952 and spread throughout Europe, reaching the United Kingdom in 1953. Wherever the virus spread, the rabbit population was devastated.

As remarkable as these early results were, they were not to last. The most lethal virus strains were so good at killing rabbits they wiped themselves out too, so after a while weaker viruses predominated as they were the ones that could go on getting passed from one rabbit to the next. The rabbits had more resistance to these weaker viruses, so because of the way that natural selection works the feral rabbits in Australia survived.

Even though the release of the *Myxoma* virus was not a 100 percent success it is still the only example of a biocontrol agent having a major impact on the population of a vertebrate. Other pathogens have been investigated as biocontrol agents to build on the *Myxoma* results, one of which is the rabbit calicivirus, a pathogen with an unknown origin that first appeared in China and which causes rabbit hemorrhagic disease. This virus escaped into the Australian rabbit population in 1995 after being tested off the coast and since then it has been better at controlling rabbits in wetter habitats, rather than dry habitats, the reverse being true for the *Myxoma* virus. Because no one is certain of the origins of this virus, concerns have been raised that it may be able to jump species, possibly even to humans.

FURTHER READING

Fenner, F., B. Fantini, and B. Fantoni. *Biological Control of Vertebrate Pests: History of Myxomatosis—An Experiment in Evolution.* CABI, Oxford, United Kingdom, 1999.

Pimental, D. *Encyclopedia of Pest Management.* CRC Press, Boca Raton, LA, 2002.

Feral Goats

The humble goat, *Capra hircus,* descends from a mountain animal native to the western highlands of Iran, a beast that was domesticated by our ancestors some 10,000 years ago. The goat appealed to nascent agriculturalists for many reasons: it is an extremely hardy animal; it can thrive on food that many other ungulates would find unpalatable; female goats produce copious quantities of nutritious milk; and when the animal has reached the end of its useful life it can be slaughtered for its meat, hide, and fur.

The goat is a superb domesticated animal; however, some of the attributes of this animal so admired by farmers are also the reason why it is somewhat of a pest in many places around the world. Humans are squarely to blame for goats becoming pests because it is our forebears who introduced this animal to locations far outside its native range. In areas lacking large herbivorous ungulates, such as Australasia and isolated oceanic islands, the introduction of goats has had far-reaching consequences

Able to survive on a very meager diet, goats are tough animals that can thrive almost anywhere. (Albert Sim | Dreamstime.com)

for the flora and fauna of these lands, much of which is endemic. It is the hardiness of goats that allows them to thrive quite happily without human intervention in areas where they have been introduced. In this sense they are the perfect animal for colonists: low maintenance and a source of much-needed animal protein and fat. Inevitably, goats often just wander off and establish feral populations and in locations where competition and predators are lacking, their populations can explode.

The ability of goats to thrive on a huge range of plants has been well documented. The goat population of Auckland Island off the coast of New Zealand ate woody plants, grasses, herbaceous plants, ferns, and seaweeds—amounting to around 40 plant species in total. The unique anatomy of the goat allows it to process such a wide range of plant food, much of which is completely off the menu for other domesticated animals. Like the other even-toed ungulates, goats are ruminants and digestion of the food they eat takes place in the anterior reaches of their digestive tract, aided by a variety of symbiotic bacteria. Rumination allows efficient digestion of vegetation, enhanced in the goat by a relatively large gut volume for the mixing of ingested food and the digestive juices suffused with bacteria.

Guadalupe, an island about 240 kilometers off the Pacific coast of Mexico, perfectly exemplifies how goats and their superefficient herbivory in the wrong place can spell disaster for native wildlife. Goats were introduced to Guadalupe at some point in the 19th century by Russian whalers and hunters who came to catch and kill sea otters, fur seals, and elephant seals. On their ships they carried goats as a source of meat and milk, and as a way of caching supplies on their voyages they left some goats on Guadalupe. The idea was that the goats would survive and the whalers could pick up some fresh meat and milk the next time they were passing. Not only did the goats survive, but they bred in profusion and before long there were at least 100,000 of them running riot over the once virginal Guadalupe, an island about 35 kilometers long and 9.5 kilometers wide. Goats consume a huge range of vegetation with almost mechanical efficiency and the verdant habitats of this Pacific island stood little chance. Before the introduction of the goat to Guadalupe there were extensive forests, but at the peak of the goat invasion much of this was munched out of existence, leaving tiny pockets of trees and closely cropped chaparral with swaths of bare ground. With the constant goat onslaught, natural regeneration of the island's vegetation was halted as any seedlings were rapidly gobbled up by these animals. Not only are goats capable of wiping out native vegetation, but the delicate ecosystems of these isolated lands, once denuded of their vegetation, are very susceptible to erosion and the plant-sustaining soil—an accumulation of hundreds of thousands of years of weathering and decay—is dried out and eroded by the wind and rain, exacerbating the action of the goat's incessant jaws.

It is only in recent years that the ecological importance of many areas with goat problems has been recognized. Populated by rare, often endemic flora and fauna, these isolated land masses are microcosms of evolution and their importance in global biodiversity cannot be overstated. The islands of the Galapagos archipelago and their unique denizens, instrumental in stimulating Charles Darwin's formulation of the theory we know today as evolution, are plagued by a burgeoning population of feral goats that threaten to undermine their fragile ecosystems. In many places around the world the feral goats, through no fault of their own, have become the target of wholesale slaughter. To date, feral goats have been eradicated from around 120 islands around the world, at great cost. These islands are often very distant from the mainland, so the expense of getting people and equipment to these areas over extended periods of time can rapidly stack up. Techniques used to control and eradicate goat populations include hunting from the land and air, radio-tracking, and poisoning. Depending on the size of the goat population and the area in question an eradication program may take many months, even years, and in some places such as Australia, the feral goat population exists over such a huge area that eradication is a practical impossibility. It is estimated that the feral goat problem in Australia is responsible for losses of around $25 million every year, a sum that includes agricultural losses and the expense of control, but which does not take into account the impact on native flora and fauna.

Like the cane toad and so many other animals derided as pests, the feral goat problem is of human origin. Up until relatively recently, the ecological consequences of moving species around the planet, purposefully or otherwise, was ignored or simply not understood. The relationship between all the organisms in any given ecosystem is a finely balanced dynamic. Introducing or removing a species into or from this dynamic has consequences for the ecosystem as a whole and we have slowly come to realize the error of our past ways.

FURTHER READING

Campbell, K. J., and C. J. Donlan. A review of feral goat eradication on islands. *Conservation Biology* 19(5)(2005): 1362–74.

Coblentz, B. E. The effects of feral goats (*Capra hircus*) on island ecosystems. *Biological Conservation* 13(4)(1978): 279–86.

Feral Pigeons

Pigeons are as much a feature of urban life for many people as air pollution, traffic jams, and crime. These widespread birds have an interesting heritage. Their ancestor is the rock pigeon, *Columba livida,* a pigeon that can still be seen in its natural haunts today: the cliff faces and rocky escarpments throughout Europe, North Africa, and western Asia. Rock pigeons were first domesticated at least 5,000 years ago, initially, it seems, for religious rites, but it is likely they were used for food too. Over time,

Feral pigeons, descendants of the rock pigeon, are found in cities and towns around the world. (USDA)

some of these domesticated pigeons escaped and took to living ferally in and around human settlements. These are the feral pigeons we know today and to them the stone, concrete, and metal structures of our cities are akin to the ledges and cliff faces of their ancestral home, which is the simple reason for their success in urban environments.

The feral pigeon is found around the world, often thriving in huge numbers in urban areas. Globally there are hundreds of millions of feral pigeons, possibly even billions, living in cities and towns. Like their forebears, the rock pigeons, feral pigeons make use of any suitable ledge or crevice for the purposes of breeding. In a sheltered spot they build a very rudimentary nest, nothing more than a messy platform of twigs and other material onto which two white eggs are laid. The male and female pigeon take it in turns to brood the eggs. The hatchlings, when they break free of their eggs, are reared on an interesting substance produced by the adult birds, which is known as crop milk. This is a not a milk in the mammalian sense, but a secretion from the lining of the bird's crop, the muscular sac at the anterior end of the digestive tract found in all birds. This secretion is similar to cottage cheese in consistency and it is very high in protein and lipids, which are crucial to the developing young. On this nutritious diet the young pigeons grow rapidly and in around 30 days they are ready to fledge the nest and strike out on their own. With this rapid rate of development pigeons are able to produce six broods every year in optimum conditions. Not only are feral pigeons able to breed everywhere in our cities, producing large numbers of young, but they are admirable opportunists when it comes to food. They'll eat just about anything, from their natural diet of seeds and berries to scraps of food dropped by humans, edible refuse, and invertebrates. In many cities it's a common sight to see a pigeon wrestling with the remnants of a burger bun, throwing it around while trying to dislodge beak-sized morsels to swallow.

Needless to say, the feral pigeon's liking for our cities has made it many enemies among those people who seek to keep our urban areas clean and animal-free. These people consider the feral pigeon to be on a par with those unsavory urbanites, the rats and mice. Indeed, feral pigeons are often referred to as "rats with wings," an inventive, albeit undeserved, moniker. Urban authorities have a problem with the pigeon for a number of reasons. The copious droppings they produce deface buildings and are thought to be a public health menace. There is no questioning the detriment to architectural aesthetics caused by the smears of accumulated pigeon droppings, and for those buildings constructed from sedimentary stone, the acidic nature of this material can exacerbate the process of erosion, an issue of particular pertinence for buildings and monuments of historical significance. Large populations of feral pigeons in urban environments can also be a problem around airports because air strikes involving pigeons occur as planes take off.

Much has been written about the danger posed to public health by feral pigeons. It is true these animals are known to harbor organisms that cause disease in humans: 60 different pathogens in actual fact. However, only seven of these are known to be transmitted to humans and of these only two pose any real risk: the bacteria *Chlamydophila psittaci,* which causes psittacosis, and the yeast-like fungus, *Cryptococcus neoformans,* the causative agent of cryptococcosis. Both of these diseases can be very serious. With that said, between 1941 and 2003 there have been only 176 documented cases of feral pigeon-borne diseases in humans. Of these cases, 99.4 percent have involved aerosol transmission, where the pathogen has been stirred up into the air from dried pigeon droppings. In healthy humans, even those with regular exposure to pigeons and their dried droppings, the risk of contracting a feral-pigeon borne disease is very low indeed. The risk of dying from a bee sting is far, far greater. Healthy people have very little to fear from feral pigeons, but there are many people around the world with diseases or conditions, an example of which is AIDS, that impair the ability of the body to fight off infection. The risk of acquiring pigeon-borne diseases from dried droppings of these birds is around 1,000-fold higher in patients whose immune system is suppressed, compared with healthy individuals. *Cryptococcus neoformans* in particular can go on to cause fungal meningitis in 2–30 percent of patients with AIDS, a disease with a very poor prognosis.

In addition to the mess caused by large populations of feral pigeons and the numerous microorganisms they harbor, these birds also play host to a number of larger parasites, specifically insects and mites, some of which are capable of biting humans or eliciting immune reactions. The most important of these are the red blood mite, *Dermanyssus gallinae,* and the pigeon tick, *Argas reflexus.* The former is responsible for nothing more than irritating bites, but repeated bites from the pigeon tick can trigger potentially fatal anaphylactic shock. Humans are also known to become sensitized to the various antigens shed by the pigeons themselves, including particles of skin, feathers, or dried droppings that are disturbed and aerosolized.

Beyond the public health and urban sanitization issues associated with pigeons, these birds can also be a problem for farmers as they have a fondness for seeds and grain, either when it has been sown or when it is ripe for harvest. With their rapid rate of metabolism, feral pigeons need a lot of food and a single bird can consume as much as 28 kilograms of food in a single year, an appetite that puts them at odds with anyone growing plants for pleasure or profit. Feral pigeons are also a problem as invasive species as they have been introduced inadvertently or otherwise to a number of isolated islands where they can be detrimental to the native fauna, especially other bird species. They compete with these native animals for food and nesting sites in some situations, but a more pervasive threat is the transmission of pathogens, including the protozoan, *Trichomonas gallinae,* which causes the potentially fatal Newcastle disease. This pathogen is responsible for the deaths of many endemic birds in the Galápagos Islands.

Even though the pest status of pigeons has been somewhat inflated, hundreds of millions of dollars are spent each year around the world trying to control their numbers as well as cleaning up the considerable mess that they leave in their wake. In the United States, damage caused by these birds in urban areas alone is estimated to cost around $1.1 billion every year. Controlling feral pigeons is very difficult because it is not as though they are predominantly crop pests, spending much if not all of their time away from centers of human population. They live among us, so broadcast spraying of pesticides specifically intended to kill birds is out of the question as these compounds are also toxic to humans, although fumigation is sometimes used in small, relatively enclosed spaces. Poisoned baits can sometimes work, but then authorities are faced with a public outcry and dealing with piles of dead pigeons.

The usual strategy for countering the feral pigeon problem in the world's cities is to minimize the areas available to them for roosting and breeding. Buildings with lots of recesses and ledges where the pigeons can roost and build their nests can be surrounded with fine netting, denying the birds access. In other areas strips of metal spikes can be attached to the sensitive parts of a building to prevent the birds from landing. Still other methods to deter them from settling on buildings include strong, long-lasting adhesives that snare the birds—a moderately successful method that is deemed to be cruel because the birds often manage to wrench themselves free, leaving toes or an entire foot behind in the process. Some authorities opt for more natural means and employ raptor handlers to fly birds such as peregrine falcons around buildings, forcing the local feral pigeons to scatter for their lives. In some places, the pigeons have become wise to this, learning the falcon is not searching for prey. Other novel methods for suppressing feral pigeon populations are the use of baits laced with contraceptives to curb the bird's ability to crank out young, and strategic placement of large nest boxes where the birds are encouraged to nest and lay eggs. The eggs, once laid, are removed and disposed of, so helping to control the pigeon population.

FURTHER READING

Haag-Wackernagel, D. Parasites from feral pigeons as a health hazard for humans. *Ann Appl Biology* (2005): 147, 203.

Haag-Wackernagel, D., and H. Moch. Health hazards posed by feral pigeons. *J. Infect.* 48(2004): 307–13.

Johnston, R. F., and M. Janiga. *Feral Pigeons.* Oxford University Press, Oxford, United Kingdom, 1995.

House Mouse

Like its larger relatives, the rats, the house mouse is a ubiquitous rodent—a small, wary animal that has taken full advantage of the opportunities offered by human settlements. The origins of this rodent are a bone of contention, but it is thought to be a native of Asia that started associating with humans when we forsook our hunter-gatherer ways for a more settled, agricultural existence. Some experts suggest this relationship first developed in northern India at least 10,000 years ago when the wild ancestors of the house mouse found the stores of food made by the first agriculturalists to their liking and began spending more and more time in and around these settlements feeding on the abundant food available.

When the settled, agricultural way of life began to spread, it was accompanied by the house mouse, an expansion that was provided with considerable impetus when global exploration began in earnest. Today, there are few places without house mice. There are even isolated, oceanic outposts, such as the Marion Island in the Indian Ocean, with a thriving population of these rodents. Up until some point after the American War of Independence, North America is thought to have been free of this animal, but stowaways on transatlantic shipping soon changed this and today the house mouse is ubiquitously distributed throughout the continent.

The house mouse, like all the rodents that live in close association with humans, is a prolific breeder, able to produce young almost like a conveyor belt. Gestation is a mere 18–20 days with each litter containing up

House mice can produce huge numbers of offspring, allowing their populations to grow very quickly when conditions are favorable. (Wotan | Dreamstime.com)

to 13 young, although 4–7 young is more normal. On average, a female house mouse can give birth to eight litters every year and the typical life span in the wild is around 2 years. With such fecundity it is no surprise that the populations of this rodent are able to reach immense sizes in a short period of time. In Australia, which suffers from intermittent house mice plagues, densities can reach more than 1,000 animals per hectare, although 10 individuals per hectare is more commonly seen.

Mainly nocturnal animals, house mice have excellent senses of smell, hearing, taste, and touch, but relatively poor eyesight. Naturally wary animals, they leave their daytime refuges to search for food, which can be just about anything. Their natural preferences are seeds and grain, but any food high in sugars, fats, or proteins will be nibbled. Interestingly, and in contrast to their larger relatives, the house mouse can survive with little or no free water, as its minimal moisture requirements appear to be met by the food it eats. The daytime shelter of a house mouse is a rough ball of shredded, fibrous material, such as paper or whatever else is easily available, and this ball can be situated underground or in any suitable, inaccessible space indoors.

House mice are a problem for a number of reasons. They eat food intended for humans and domesticated animals and at the same time they spoil even more food with their feces and malodorous urine and skin secretions. Typically, they nibble at food, damaging and contaminating a wide range of foodstuffs, rather than consuming any one item in its entirety. In a 12-month period, a single mouse can consume about two kilograms of food and produce around 18,000 droppings. The small size of the house mouse means that it has a large surface area in relation to its volume; therefore, it loses a lot more heat than a larger animal and to compensate for this loss it has to consume a large proportion of its body weight every day—around 10–15 percent. Crop and stored food losses traceable to house mice must be immense, but are very difficult to estimate accurately. In some areas it has been estimated that mouse damage can cause 50 percent preharvest losses in crops as diverse as cereals, legumes, pulses, sorghum, maize, peas, beans, and chickpeas, zucchini, tomatoes, eggplants, capsicums, and melons. When this preharvest damage is combined with losses incurred by mouse feeding and contamination postharvest, the global economic impact must be immense—probably many billions of dollars every year. One plague of house mice in Australia in 1993 was estimated to cost the agricultural industry at least 100 million Australian dollars, both in terms of crop losses and the expense of controlling the outbreak.

The copious feces and urine mice produce are also laden with a diverse fauna of microorganisms, many of which can cause disease in humans and animals. Potentially the most dangerous bacteria transmitted by these mice are those in the genus *Salmonella,* pathogens that can cause serious cases of food poisoning in humans, which can frequently be fatal in young, elderly, or sick individuals. Other organisms transmitted by house mice include the small tapeworms, *Hymenolepis nana* and *H. diminuta,* the eggs of which can be inadvertently ingested by humans if the hands, food, or water are contaminated by mouse feces.

Mice are not known for their aggression, but they can defend themselves with their large incisors, and rat-bite-like fever can be caused by bites from these rodents. Weil's disease can also be transmitted in food or water contaminated with house mouse urine. A fungal disease of the scalp known as favus is also transmitted by these rodents, either via direct contact with the mice themselves or indirectly via cats. House mice are also a reservoir for the pathogens that cause plague and murine typhus, diseases that are transmitted via the bites of fleas. A mite that lives on house mice, *Liponyssoides sanguineus,* is known to transmit rickettsial pox. Lymphocytic choriomeningitis and poliomyelitis (potentially) can be transmitted to humans via house mouse feces. House mouse mites are also known to cause dermatitis when they feed on humans.

The other major problem with house mice is the damage they cause to property. In finding suitable places to nest and feed they will quite happily gnaw through a range of different materials, including electrical wiring, increasing the risk of fires. In areas that support intermittently enormous mouse populations, tunneling activities can undermine buildings and other structures.

Finally, the house mouse is considered a pest because of the impact it can have on native flora and fauna, which can be particularly disastrous when these animals find their way to oceanic islands where the wildlife has evolved in the absence of a small opportunistic rodent. A perfect example of how destructive house mice can be as invasive species is on Gough Island, a small isolated outpost in the South Atlantic that supports a huge population of nesting seabirds and is also home to two endemic land birds (Gough moorhen and Gough bunting). The mice on Gough Island were accidentally transported there around 150 years ago by British ships and the lack of predators allowed their population to explode. In a severe error of judgment it was decided to introduce cats to the island in order to control the burgeoning mouse population, but the cats soon learned the native birds were much easier prey than the wary mice and the populations of many bird species nose-dived, prompting the extermination of the cats. In the absence of the cats, a worrying trend is developing among the numerous mice on the island. They are growing larger and they are feeding on sea bird chicks, which are left for extended periods of time while their parents are out at sea finding food. The mice gather around nestling albatross chicks at night and nibble away at their flanks and underside. The resultant blood loss is enough to severely weaken and even kill the young sea birds. The opportunistic feeding activity of the mice on Gough Island is thought to be contributing to declines in breeding success among some of the sea birds that nest on this island.

House mice, like so many other pests, will never be eradicated. They are simply too widespread and adaptable to be wiped out by any of the practical control measures that humans can throw at them. Like their larger relatives, the rats, house mice are survivors. Over the last few thousand years, evolution has honed the house mouse to take advantage of whatever opportunities humanity has presented it. Regardless of its supreme adaptability there are some simple ways of controlling the populations of these animals by limiting the opportunities available to them in terms of hiding/nesting sites and food resources. Any areas routinely used by house mice for nesting and routes used by the animals to move from one area to another should be blocked up or destroyed. Food should always be safely secured and any food waste should promptly be cleaned up to avoid attracting mice. Rondenticides are chemicals used primarily to kill mice and their relatives and they are normally used to lace food-baits that mice find tempting. These poison-laced foods can be left in small bait stations positioned against walls that mice run along to get from place to place. There are also myriad traps to catch and kill mice, including the archetypal mousetrap, as well as some humane traps that allow the mouse to be caught and released where it won't cause a problem. Contrary to popular belief, cats are not that good at exterminating mice, especially if there's an established infestation. They will catch and kill a few mice, but the vast majority of these wary animals will evade the cat and continue to feed and breed unabated. Cats and other predators can help to stop mice from reinfesting an area that has been cleared, as mice will be deterred from entering buildings by the distinct smell of any of their many predators.

The house mouse, as loathed as it is, is a born survivor—an animal that is perfectly suited to surviving and thriving in an increasingly human-dominated world. Perhaps the best strategy for dealing with this pest is to recognize the ways in which they can harm us and limit these as much as possible, rather than by spending billions of dollars every year trying to rid our lives of them.

FURTHER READING

Buckle, A. P., and R. H. Smith (eds.). *Rodent Pests and Their Control.* Oxford University Press, Oxford, United Kingdom, 1996.

Cuthbert, R., and G. Hilton. Introduced house mice *Mus musculus:* A significant predator of threatened and endemic birds on Gough Island, South Atlantic Ocean? *Biological Conservation* 117(2004): 483–89.

Jones, A. G., S. L. Chown, and K. J. Gaston. Introduced house mice as a conservation concern on Gough Island. *Biodiversity and Conservation* 12(2003): 2107–119.

Meehan, A. P. *Rats and Mice: Their Biology and Control.* Rentokil, East Grinstead, United Kingdom, 1984.

Red-billed Quelea

Generally, birds are not important pests of agriculture. There are those bird species that can sometimes be a nuisance for farmers at specific times of the year, but on the whole, birds are relatively benign. The notable exception to this is the red-billed quelea (*Quelea quelea*) of Africa, a small weaver bird widely considered to be the most numerous of wild birds on the planet.

In its native Africa, this small bird naturally feeds on the seeds of wild grasses; however, when its preferred food is scarce, the red-billed quelea quite happily turns its attentions to crops, specifically the seeds of millet, sorghum, rice, barley, and wheat. In many areas, the red-billed quelea is known as the locust bird for the devastation it can cause in agricultural crops. The main problem with this bird is its huge population. A single, sky-blackening flock of red-billed quelea can contain more than 1 million birds and as birds are very active animals they require a lot of food to keep their bodies functioning. A single red-billed quelea weighs around 20 grams and it needs around half of its body weight in food every day;

An adult male red-billed quelea. This is probably the most numerous bird on the planet. (Linncurrie | Dreamstime.com)

therefore, a flock of 1 million of these birds can consume 10 tonnes of grain every day—grain intended for the mouths of poor subsistence farmers and their families. Africa is thought to be home to at least 1.5 billion red-billed quelea, so the entire population of this bird can consume a considerable amount of food every year. The economic losses due to red-billed quelea are estimated to be at least $50 million annually, but the real figure may be far higher.

Needless to say, humans can't find it within themselves to tolerate this sort of competition, even if it is from a bird. In helping themselves to our food, red-billed quelea are squarely in the firing line of most farmers in sub-Saharan Africa and to date just about every technique for killing and destroying has been thrown at these birds without much success. Being birds, red-billed quelea are able to migrate large distances on the lookout for food resources and they also breed very rapidly, producing three clutches of eggs every year, each of which contains around three eggs. Winged, wary, and fast breeders, red-billed quelea are very difficult to control. In South Africa alone, millions of these birds are killed every year in a variety of ways, but the wholesale slaughter has done nothing to dent their numbers.

Farmers and pest controllers in Africa have been creative in coming up with ways to bring about the demise of their feathered foes. They poison them, blow them up, set fire to them, and generally make things very difficult for the red-billed quelea. From the bird's perspective it must seem as though all-out war has been declared. Poisoning these birds involves the use of fenthion—commonly known as quelea-tox (an organothiophosphate) and alpha-chloralose—both of which are sprayed from aircraft above dense aggregations of the birds when they're feeding or nesting. Fenthion is a neurotoxin and it kills the birds by interfering with the way in which nerve impulses are transmitted across chemical synapses. This can't be a nice way to go for the birds. Furthermore, fenthion is also toxic to many other organisms, including other birds, terrestrial invertebrates, and aquatic creatures, so huge numbers of organisms will be killed in the area over which the compound is sprayed. Alpha-chloralose immobilizes the birds and kills them. Luckily for the environment, fenthion and alpha-chloralose and the equipment needed for their large-scale application are beyond the means of most farmers in sub-Saharan Africa, so they resort to more primitive ways of controlling the birds. Red-billed quelea are weaver birds and they brood their eggs and rear their young in intricate hanging nests constructed from grasses and other plant material. The birds make their nests in huge colonies that can cover an area equivalent to several football pitches. It is these nesting colonies that are targeted by farmers, as the red-billed quelea are there at their most vulnerable. A common means of destroying the nesting quelea is by placing explosives below the nest trees and detonating them. For all the noise and destruction these explosives cause, their impact on the red-billed quelea populations is negligible. The farmers also set fire to the areas in which the birds nest and use flamethrowers to burn their nests.

Less destructive ways of controlling the red-billed quelea population include scaring them from crops with loud noises when they alight to feed. This can be moderately successful in the short term, but the birds soon become accustomed to the disturbances and carry on feeding regardless. There are also plans afoot to use the red-billed quelea as a source of dietary protein for people in sub-Saharan Africa. Protein is severely lacking in the diets of many subsistence farmers and sustainably harvesting the red-billed quelea may kill two birds with one stone. It remains to be seen if this proposal is practically possible and if it would be widely accepted by subsistence farmers throughout sub-Saharan Africa.

The methods described above kill millions of red-billed quelea every year. In South Africa between 1995 and 1998, chemicals accounted for the deaths of almost 28 million of these birds at a cost of around $130,000. During the same period, explosives killed just over 50 million red-billed quelea at a cost of around $225,000. With a slaughter of this intensity it is hard to believe that any of these birds are left at all; however, at the moment, the red-billed quelea problem continues unabated. As the cultivation of the sub-Saharan landscape increases, the amount of food available to the red-billed quelea will also increase and the populations of this bird will grow. With this said, there will come a time, if human population growth continues, where much of sub-Saharan Africa will be under the plough and the natural landscapes that epitomize this continent will be lost along with the nesting sites used by these birds. Africa is certainly on a trajectory that will see it lose its singular biodiversity and the problems we see today with red-billed quelea will be nothing but a distant memory.

FURTHER READING

Briggers, R. L., and C.C.H. Elliot. *Quelea quelea: Africa's Bird Pest.* Oxford University Press, Oxford, United Kingdom, 1990.

Sea Lamprey

The sea lamprey is a primitive vertebrate native to the Atlantic Ocean. Superficially, these animals resemble eels, but fundamentally they are very different creatures. For one thing, instead of jaws they have a circular disk for a mouth and a rasp-like tongue wreathed by concentric circles of small curved teeth. Essentially a parasite as an adult, the lamprey uses the nightmarish oral disc to latch on to a suitable fish and rasp at its flesh. Also, instead of a calcified skeleton it has an internal scaffold composed only of cartilage. In addition it has no lateral line (a sense organ found in bony, jawed fish), no vertebrae, no swim bladder, and no paired fins.

Lampreys are bizarre animals in both behavior and lifestyle. The sea lamprey's life cycle begins with the adults ascending tributary streams, constructing nests, and spawning in the gravel beds of these streams. The 30,000–100,000 eggs the fertilized females deposit hatch and the small, worm-like larvae, very different in appearance from the adults, get swept downstream. The young end up burrowing into sand and silt, where they filter edible matter from the water. After 3–17 years of this sedentary, filter-feeding lifestyle, the larvae, now around 15 centimeters long, transform into the parasitic adult in the late summer/early fall and leave their burrow for life in the open water of the ocean or, nowadays, the Great Lakes. During this transformation, the lampreys develop eyes, and their distinctive vicious-looking oral disk and their kidneys undergo changes that in their natural range would allow them to return to the saline environment of the open ocean. The adults go about feeding from any suitably sized fish they can latch onto and once they're sexually mature the life cycle goes full circle with the females returning to the streams to lay eggs. The entire sea lamprey life cycle takes six years on average, although it can take as long as 20 years. Fully grown, a female sea lamprey can be as much 90 centimeters long and 2.5 kilograms in weight.

As has already been mentioned, lampreys feed parasitically on other fish. They use their numerous, curved teeth to grab onto the flank of a passing fish and then their tongue rasps away the flesh to feed on the tissue, blood, and other bodily fluids. To prevent the blood from clotting during feeding, the lampreys produce an anticoagulant. When the lamprey has finished feeding, the unfortunate victim is left with a raw, disk-shaped wound that permits the entry of pathogenic organisms. The blood loss, infection of the wound, and entry of other disease-causing organisms can severely weaken the victim, commonly resulting in death.

The head of an adult sea lamprey. (Lee Emery, US Fish and Wildlife Service, Bugwood.org)

In their natural habitats, lampreys are simply another component in the ecosystem; however the sea lamprey, aided by humans, has found its way into the Great Lakes of America from its natural home in the Atlantic Ocean. Prior to the advent of large-scale shipping the Niagara Falls in North America was a natural barrier to oceanic fishes such as the sea lamprey. However, to connect the industries that were springing up around the shores of the Great Lakes with markets around the world, a series of locks and canals to guide water vessels around these natural barriers was constructed.

Following the completion of these conduits, shipping was freely able to ply the route between the Great Lakes and the Atlantic. So, too, were various invasive species and among the most destructive of these was the sea lamprey. Sea lampreys were first spotted in Lake Ontario in the 1830s and by 1938 they had reached Lake Superior. The absence of natural enemies in the Great Lakes allowed the sea lamprey population to explode and by the 1940s this species was devastating the populations of native fish, such as lake trout, salmon, rainbow trout (steelhead), whitefish, chub, burbot, walleye, and catfish. Each sea lamprey in its lifetime can destroy about 18 kilograms of fish that were the mainstay of the Great Lakes fishery. Before the spread of the sea lamprey, more than 6,000 tons of lake trout alone were landed every year in the Great Lakes. In the early 1960s, following the meteoric spread of the lamprey, the annual commercial catch of lake trout from these lakes had fallen to around 130 tons.

A considerable amount of time and money has been spent over the last six decades in an effort to try and control this invasive species. To date, these measures have achieved a good degree of success. Although the sea lamprey population exploded following its spread through the Great Lakes, the species does have certain characteristics that render it vulnerable to control measures. First and foremost is the fact that sea lamprey require spawning streams with specific features. Currently, of the 5,747 streams and tributaries that feed the Great Lakes, 433 are known to be used as spawning grounds by the sea lamprey. Preventing the adult lampreys from entering these streams and spawning will obviously reduce the population of this invasive parasite. Various types of barrier, weir, and trap have been constructed in the mouths of these tributaries, some of which have been more successful than others, which have suffered problems with maintenance. The other vulnerable part of the sea lamprey's life cycle is that of the sedentary larva, which has been targeted with compounds known as lampricides. During the 1960s more than 6,000 compounds were screened by the U.S. fish and wildlife service, with the eventual identification of one compound, TFM (3-trifluoromethyl-4-nitrophenol). This compound was found to be selectively toxic to sea lampreys, specifically the vulnerable larvae when they are in their burrows in the spawning streams. On a four-year rotation schedule TFM is used to treat around 250 streams in the Great Lakes area. The four-year rotation allows the populations of certain aquatic invertebrates to recover in between treatments. Although the current consensus is that this compound is about as environmentally neutral as a synthetic toxin can be, it remains to be seen what the long-term consequences of continued application of this compound will be for the Great Lakes ecosystem.

In view of concerns over the unforeseeable consequences of introducing considerable quantities of TFM into lamprey spawning streams each year, scientists are investigating other means of controlling this invasive species to reduce reliance on chemical control. These experimental methods include sterile male release and pheromone traps, both of which could replace or reduce the need for TFM applications.

The measures described above have succeeded in reducing sea lamprey populations by around 90 percent, which is nothing short of a remarkable success. With the sea lamprey population now under control, the Great Lakes commercial and sport fishery has begun to recover. However, the sea lamprey will never be eradicated from the Great Lakes and it is important to remember that human activities are to blame for this fishy problem. The requirements of commerce saw the development of the canals and locks that linked the previously isolated Great Lakes to the open ocean, opening this freshwater ecosystem to a large number of invasive species. These invading organisms have proved to be an expensive, perennial problem for North America, both environmentally and economically.

FURTHER READING

Hardisty, M. S., and I. C. Potter. *The Biology of Lampreys.* Academic Press, London, 1982.

Snakes

Snakes are fascinating animals with a fearsome reputation. Represented by approximately 3,000 species around the world, these carnivorous reptiles have lost their legs and with no limbs to hold or subdue prey they have evolved some remarkable ways of feeding. Most snakes, around 2,400 species, subdue their prey by using their long muscular bodies like a vise to squeeze the life out of their unfortunate victims, which can range in size from the tiny young of a rodent to large, predatory mammals, even humans. The minority of snakes, approximately 600 species, use a different and extremely successful strategy for capturing their prey, a technique that hinges on saliva that has been shaped by evolution into a complex toxin—venom—which the snake injects into its victim using modified teeth—fangs.

As interesting as snakes are from a purely zoological point of view, the venomous species are known to cause many thousands of deaths around the world each year, not to mention the severe debilitation that snake envenomation can inflict on those victims who survive a bite from a dangerous species. In the developed world, snakebites are not really a problem and a fatal envenomation is something of a rarity. The situation in developing countries is rather different and as human populations continue to encroach on snake habitat the problem of snake envenomation is likely to worsen.

Cobras are responsible for many deaths each year, especially on the Indian subcontinent. (Vishwa Kiran | Dreamstime.com)

Defining how big the snakebite problem is has always been difficult, because it can only be estimated. Victims in remote areas may die without the relevant information ever reaching the authorities and there may be a significant number of victims who prefer to seek traditional snakebite remedies rather than visiting a conventional hospital. Recent estimates for the number of snakebites that occur around the world every year are in the region of 1.2 to 5.5 million, of which 420,000–1.8 million involve the delivery of venom, often called *wet* bites (many snake bites are dry: the animal bites, but no venom is injected). These envenomations are estimated to cause 20,000—94,000 deaths every year. The most severely affected area is the Indian subcontinent, a region with a large number of venomous snake species and a burgeoning human population that is making increasing demands on the environment—a scenario that brings snakes and humans into direct contact with one another.

The snake problem is so large because these reptiles are successful animals that are able to survive in a range of habitats, often completely out of sight. When humans and agriculture move into an area, snakes will often take advantage of the situation to feed on the animals that are to be found wherever there are people, namely rodents. Most snakes are not aggressive animals. They don't bite casually, for the simple reason that venom is biologically expensive to produce and biting for anything other than hunting or defense would be a waste of a precious resource. In the vast majority of situations a snake will detect the approach of a human and slip off into the undergrowth or into a retreat. It's only when a snake is surprised or cornered that a bite involving the injection of venom is likely, and even then the reptile may not inject all the venom it is capable of delivering.

Most of the snakebite deaths that occur each year are inflicted by a relatively small number of species (see sidebar). These particular species' potential to cause human injury and death is high because they commonly come into contact with humans, they produce potent venom (although by no means are they the most venomous species), and they have a propensity for standing their ground and striking at a perceived threat. The Indian subcontinent is something of a microcosm of the snakebite problem because it is here that these reptiles cause the most fatalities. The reason is that a huge and rapidly growing human population is in the presence of four species of very venomous snake, which are actually responsible for the majority of all snakebite-related deaths. These are Russell's viper, the saw-scaled viper, the common krait, and the spectacled cobra. At this point, it is worth looking at the two major types of venomous snake.

Russell's viper and the saw-scaled viper belong to the most evolutionarily advanced group of snakes—the vipers. These snakes deliver venom through huge, hinged fangs. Viper venom contains many different substances, but the overall effect of envenomation by most viper species is the breakdown of proteins (proteolysis), which results in tissue destruction. Depending on the potency of the venom, how much is injected, and where it's injected on the body, the result of a viper bite can range from localized pain and swelling to disruption of blood clotting mechanisms, multiple organ failure, and death. In those victims who survive a bite from one of the dangerous vipers the tissue damage can be so extensive that amputation is the only option. Most of the people who get bitten by these snakes are poor people in rural areas who rely on manual labor for subsistence, so it's easy to understand how the loss of a limb can have devastating consequences for the lives of these people and their families.

The common krait and the spectacled cobra belong to the other very important group of venomous snakes—the elapids. The venom of these snakes is delivered through fixed fangs and in most cases the toxic mixture contains compounds that interrupt the relay of electrical impulses in the nervous

The Snake Species Responsible for the Most Human Deaths, the Potency of Their Venom, and Their Geographic Distribution

Species	Toxicity of venom (LD/50–mg/kg)*		Mean venom yield (mg)	Geographic distribution
	Subcutaneous	Intravenous		
Russell's viper (*Daboia russelii*)	0.75	1.33	150	Indian subcontinent, Southeast Asia, southern China, and Taiwan
Spectacled cobra (*Naja naja*)	0.45	0.35	200	Indian subcontinent
Saw-scaled viper (*Echis carinatus*)	0.151	-	4.6	Much of Asia, from the Middle East eastwards
Common krait (*Bungarus caeruleus*)	0.365	0.169	22	Indian subcontinent
Common lance-head (*Bothrops atrox*)	22	2.835	120	South America, east of the Andes
Neotropical rattlesnake (*Crotalus durissus*)	-	1.244	20–100	Most of South America
Puff adder (*Bitis arietans*)	4.4–7.7	0.4–2.0	100–350	Africa (except the Sahara and equatorial rainforest) and south-west of the Arabian peninsula

*The toxicity of venoms is quantified with the LD/50 (lethal dose) test, which indicates how much venom it takes to kill 50 percent of the test animals, typically mice. The lower the LD/50 value, the more toxic the venom. Also, it's worth remembering that humans are much more susceptible to many venoms than mice, so the LD/50 values for humans may be much lower than those quoted in this table.

system. Some of the Australian elapids and all the sea snake species produce venom with both proteolytic and neurotoxic components, making the bites of these species potentially very serious. Thankfully we can say that envenomations from these snakes are very rare indeed. Like a bite from a dangerous viper, the result of an elapid bite depends on the snake species in question, the potency of its venom, how much venom it injects, and where on the body the venom is injected. In some cases, an elapid bite may only result in increased sweating and anxiety, but in others paralysis of the lung and heart muscles may occur, closely followed by death. Overall, elapids are probably more dangerous than vipers because their venom generally acts on the nervous system and causes paralysis, so if proper medical care is distant the patient will be lucky to survive. In contrast, the nature of viper venom means there is more time to seek professional medical help before the victim is in serious danger of losing his or her life.

Until the end of the 19th century, bites from dangerous snakes were very often fatal, but in 1895 the French doctor, Albert Calmette, developed a technique for neutralizing snake venom along the same lines as the way in which vaccines are produced. Essentially, a tiny amount of snake venom injected into a mammal will elicit the production of protein-specific antibodies—antivenom (antivenin)—that can be harvested and then used to neutralize the venom in the body of a snakebite victim. Today, antivenoms for most of the important snake species are produced in considerable quantities and in some areas the large-scale adoption and prompt use of antivenoms has rendered fatal snake bites a very rare event indeed. Nowhere is this more evident than Australia, a country that leads the way in antivenom technology because of the large number of venomous animals that are to be found there. Australia is the only continent where venomous snakes outnumber the nonvenomous species (17 of the 20 most venomous snake species are found in Australia) and it is also home to the most venomous terrestrial snake in the world—the inland taipan (*Oxyuranus microlepidotus*), whose venom is around 50 times more toxic than that of a spectacled cobra. Between 1981 and 1991 only 18 deaths were attributable to snakebites in Australia, which is very impressive considering the number of dangerous Australian snakes and the potency of their venom.

Snakes fulfill the criteria for being a pest because of what their venom can do to people who are unlucky enough to get bitten by one of these reptiles. Globally, this wouldn't be too much of a concern if fatalities or injuries from snakebites were rare events, but as the human population continues to expand, snakes and people will come into contact more and more frequently, with the former being increasingly vilified and the latter suffering debilitating injury or death. This is more a problem for the long-term survival of snakes, especially in the developing world, as all species, regardless of their potential to harm, are seen as a threat and exterminated wherever they are found.

The pest status of snakes is something that can be addressed with some very simple measures. Firstly, food and refuse that attracts rodents, the favored prey of many snake species, can be stored or discarded in such a way that it is not available to vermin. Secondly, caution should be exercised wherever there may be snakes, as there are very few of these reptiles that will bite a human without provocation (the black mamba of Africa being a notable exception). It is perfectly reasonable to suggest that snakes and humans can live side by side as long as we respect the requirements of these animals and their position in a healthy, functioning ecosystem.

FURTHER READING

Kasturiratne, A., A. Anuradhani, R. Wickremasinghe, A. Pathmeswaran, N. S. Rajitha, N. Kithsiri Gunawardena, R. P. Arunasalam, L. Savioli, D. G. Lalloo, and H. Janaka de Silva. The global burden of snakebite: A literature analysis and modelling based on regional estimates of envenoming and deaths. *PLOS Medicine* 5(11)(2007): 1591–604.

Langley, R. L. Animal-related fatalities in the United States—an update. *Wilderness Environ Med* 16(2)(2005): 67–74.

Meier, J., and J. White. *Handbook of Clinical Toxicology of Animal Venoms and Poisons.* CRC Press, Boca Raton, LA, 1995.

Glossary

Anatomy—the structure of an animal or plant, or of any of its parts, and the study thereof.

Archipelago—a group of oceanic islands.

Arthropod—any animal belonging to the phylum arthropoda, which includes insects, arachnids, crustaceans, millipedes, and centipedes.

Biodiversity—the diversity of living things on earth.

Bivalves—the mollusc class characterized by animals with a soft body protected by two shell valves, such as mussels, oysters, and scallops.

Cardiovascular disease—any disease that affects the heart and/or the circulatory system.

Cereal—a grass such as wheat, oats, or corn, the starchy grains of which are used as food.

Commensalism—of, relating to, or characterized by a symbiotic relationship in which one species is benefited while the other is unaffected.

Congener—taxonomically, organisms in the same genus.

Cosmopolitan—relating to the geographical range of an organism when it occurs in many parts of the world.

Cryptosporidiosis—the disease caused by the protozoa in the genus Cryptosporidium.

Cultivar—a variety of a plant selectively bred to accentuate certain traits.

Cultivation—the process of growing plants on arable land.

Divergence—in an evolutionary sense the process by which two or more species evolve from a common ancestor and continue along separate evolutionary trajectories.

Ecological—of or pertaining to ecology, the study of the living earth.

Ecosystem—a system formed by the interaction of a community of organisms with their environment.

Enzyme—a biological catalyst, typically a protein that increases the rate of a reaction, making all life possible.

Esophagus—the tube that connects the pharynx (throat) with the stomach.

Exoskeleton—the chitin-based exterior skeleton of arthropods that protects the animal's internal organs as well as providing points of attachment for the muscles.

Fauna—all the animals in any given habitat or ecosystem.

Fecundity—the reproductive capacity of an organism.

Fertilizer—any substance, natural or synthetic, applied to plants to increase the levels of nutrients in the soil.

Generalist—in an ecological sense, an organism that has catholic tastes in terms of food and/or habitat requirements.

Hemiptera—the order of insects that includes all the true bugs, such as aphids, plant hoppers, scale insects, mealy bugs, lice, and so forth.

Herbivore—any animal that survives by feeding on plant matter.

Immunity—a state of having sufficient biological defenses to avoid infection.

Insect—any animal in the phylum arthropoda with six legs, jointed appendages, and (very often) wings.

Intensification—with respect to agriculture, the process by which crop yields are increased with cultural and chemical means.

Invertebrate—any animal that lacks a vertebral column.

Lagomorphs—the mammalian order that includes the rabbits, hares, and pikas.

Larva (pl. larvae)—the juvenile stage in the life cycle of an insect, which metamorphoses into the adult.

Lipophilic-—any substance with an affinity for lipids (fats and oils).

Livestock—any domesticated animal that is farmed.

Metabolism—the whole range of biochemical processes that occur in any living thing.

Microbe—any microscopic organism, which includes bacteria, viruses, and single-celled fungi.

Mitochondria—the organelles in eukaryotic cells that are responsible for converting food into usable energy.

Mollusc—the phylum of animals that includes octopi, squid, bivalves, snails, and slugs.

Morphology—the form and structure of an organism considered as a whole.

Mutation—a change in the genetic sequence of an organism's genome that can cause changes in the proteins the genes code for.

Nectar—the sugary fluid produced by plants to attract pollinating animals.

Nematode—any worm-like animal in the phylum nematode.

Neuroinhibitor—any substance that inhibits the propagation of a nerve impulse.

Neurons—the cells of the nervous system that are responsible for producing and relaying electrical nerve impulse.

Neurotoxin—any substance that is toxic to nerve cells.

Neurotransmitter—any substance produced by neurons as a means of converting an electrical nerve impulse into a chemical message for modulation of the impulse.

Onchocerciasis—disease caused by the nematode Onchocerca volvulus.

Organic—relating or belonging to the class of chemical compounds having a carbon basis.

Parasite—an organism that lives in or on and takes its nourishment from another organism.

Parasitoid—any organism that spends a significant portion of its life history attached to or within a single host organism that it ultimately kills.

Pathogen—any organism that causes disease.

Pathogenic—of or relating to pathogens.

Photosynthesis—the biological process by which sunlight is used to convert water and carbon dioxide into sugars.

Phylum—the primary subdivision of a taxonomic kingdom, grouping together all classes of organisms that have the same body plan, for example, arthropoda, nematode, and so forth.

Physiology—the branch of science dealing with the functioning of organisms.

Platyhelminth—a phylum of animals including the flukes and tapeworms.

Proboscis—the slender, tubular feeding and sucking organ of certain invertebrates, including insects.

Pupa (pl. pupae)—the resting stage in the life cycle of an insect that undergoes metamorphosis where the structures of the larva are reordered into those of the adult.

Quiescence—a state of stillness or inactivity.

Raptor—a bird of prey, such as a falcon, hawk, or eagle.

Sedentary—any animal that is fixed to one spot.

Seedbed—the prepared soil that receives the seeds during the sowing of a crop.

Toxoplasmosis—the disease caused by the protozoan *Toxoplasma gondii.*

Ungulate—any mammal with hooves.

Vector—any animal that acts as a vehicle for a disease-causing organism.

Vertebrate—any animal with a vertebral column.

Yield—the amount of a crop a given area of land produces.

Selected Bibliography

Bond, C. E. *Biology of Fishes.* Saunders College Publishing, Forth Worth, TX, 1996.

Buckle, A. P., and R. H. Smith (eds.). *Rodent Pests and Their Control.* Oxford University Press, Oxford, United Kingdom, 1996.

Capinera, J. L. *Encyclopedia of Entomology,* Vol. 2. Springer, Dordrecht, 2008.

Foelix, R. F. *The Biology of Spiders* (2nd ed.). Oxford University Press, Oxford, United Kingdom, 1996.

Goddard, J. *Infectious Diseases and Arthropods.* Humana Press, Totowa, NJ, 2008.

Gullan, P. J., and P. S. Cranston. *The Insects: An Outline of Entomology.* Blackwell Science, London, 2000.

Hickman, C. P., L. S. Roberts, and A. Larson. *Integrated Principles of Zoology.* WCB Publishing, Dubuque, IA, 2006.

Hill, D. S. *The Economic Importance of Insects.* Chapman & Hall, London, 1997.

Hoelldobler, B., and E. O. Wilson. *The Ants.* Belknap Press, Cambridge, MA, 1990.

Kearn, G. C. *Leeches, Lice and Lampreys: A Natural History of Skin and Gill Parasites of Fishes.* Springer, Dordrecht, Germany, 2004.

Lane, P. and R. W. Crosskey. *Medical Insects and Arachnids.* Chapman & Hall, New York, 1993.

Lee, D. L. *The Biology of Nematodes.* Taylor & Francis, London, 2002.

Lehane, M. J. *The Biology of Blood-Sucking Insects.* Cambridge University Press, Cambridge, MA, 2005.

Macdonald, D. *The New Encyclopaedia of Mammals.* Oxford University Press, Oxford, United Kingdom, 2001.

Margulis, L., and W. H. Schwartz. *Five Kingdoms.* Freeman and Company, New York, 1998.

Mullen, G. R., and L. A. Durden (eds.). *Medical and Veterinary Entomology.* Academic Press, San Diego, CA, 2009.

Nowak, R. *Walker's Mammals of the World.* Johns Hopkins University Press, Baltimore, MD, 1999.

Pimental, D. *Encyclopedia of Pest Management.* CRC Press, Boca Raton, LA, 2002.

Roberts, L. S., and J. Janovy, Jr. *Foundations of Parasitology.* McGraw-Hill Higher Education, New York, 2008.

Ruppert, E. E., and R. D. Barnes. *Invertebrate Zoology* (6th ed.). Saunders College Publishing, Fort Worth, TX, 1994.

Service, M. W. *Medical Entomology for Students.* Cambridge University Press, Cambridge, United Kingdom, 2004.

Walter, D. E. *Mites: Ecology, Evolution and Behavior.* CABI, Wallingford, United Kingdom, 1999.

Zug, G. R., L. J. Vitt, and J. P. Caldwell. *Herpetology: An Introductory Biology of Amphibians and Reptiles.* Academic Press, San Diego, CA, 2006.

Web Resources

Centers for Disease Control and Prevention—www.cdc.gov
Food and Agriculture Organization of the United Nations—www.fao.org
European Center for Disease Prevention and Control—www.ecdc.europa.eu
European and Mediterranean Plant Protection Organization—www.eppo.org
Invasive Species Specialist Group—www.issg.org
United States Department of Agriculture—www.usda.gov
University of Florida Department of Entomology and Nematology—www.entnemdept.ufl.edu
World Health Organization—www.who.int

Index

About the Author

ROSS PIPER is an independent scholar. His lifelong interest in natural history, especially animals, led to academia and he went on to gain a first-class degree in zoology from the University of Wales, Bangor, and a PhD in entomology from the University of Leeds. Currently, he lives in Hertfordshire, England. This is his eighth book.